For Our Kids' Sake

How to Protect Your Child Against Pesticides in Food

by Anne Witte Garland

Foreword by T. Berry Brazelton, M.D.

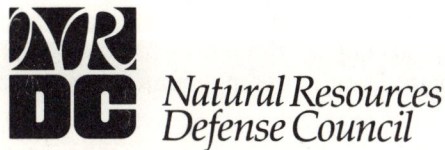

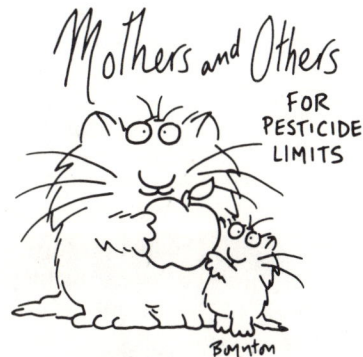

NRDC
Natural Resources
Defense Council

Mothers and Others
FOR PESTICIDE LIMITS

Boynton

A PROJECT OF THE
NATURAL RESOURCES DEFENSE COUNCIL

The Natural Resources Defense Council (NRDC) is a nonprofit membership organization dedicated to protecting natural resources and improving the quality of the human environment. With offices in New York City, Washington, D.C., and San Francisco, and a full-time staff of more than 80 lawyers, scientists, and environmental specialists, NRDC combines legal action, scientific research, and citizen education in an extremely effective environmental protection program. NRDC has more than 100,000 members and is supported by tax-deductible contributions. Since its founding in 1970, NRDC has worked to improve air quality, protect food and drinking water supplies from toxic contamination, shape more efficient national energy policies and nuclear safety regulation, and promote wise stewardship of public lands and marine resources.

Mothers and Others for Pesticide Limits, a project of NRDC, is working nationally to call attention to the problem of pesticides in children's food, to press for reforms in pesticide regulations and enforcement, and to ensure that safe produce is widely available around the country.

Natural Resources Defense Council, 40 West 20th Street, New York, New York 10011

For information on discounts for bulk orders of *For Our Kids' Sake*, write the NRDC Publications Department at the above address.

The principal authors of "Intolerable Risk: Pesticides in Our Children's Food" are Bradford H. Sewell and Robin M. Whyatt, M.P.H., with contributing authors Janet Hathaway, Esq. and Lawrie Mott, M.S.

For Our Kids' Sake was edited by Paul Allen, Jane Bloom, Esq., Cynthia Hampton, Al Meyerhoff, Esq., Lawrie Mott, M.S., Wendy Gordon Rockefeller, Meryl Streep, Stephen Tukel, and Robin M. Whyatt, M.P.H.

Production supervision by Linda Lopez, with production assistance provided by Barbara Courtney, Kent Curtis, Rennie Hutton, Jennifer Stenzel, and Marie Weinmann

Cover design and illustrations by David Garland
Book design by Anne Witte Garland
Cover photo by Bruce Wall
Inside photos U.S. Dept. of Agriculture
Typography by Brad Allan Walrod
The text of this book is printed on recycled paper.

This book was written and produced with funding by the Rockefeller Family Fund. The research and writing for "Intolerable Risk: Pesticides in Our Children's Food," the two-year NRDC study on which this book is based, was funded by The Kettering Family Foundation and The Dakin Fund, with additional support contributed by Tom and Margaretta Brokaw, Saul and Amy Cohen, and Patricia Kind. In addition, support for this book and for "Intolerable Risk" came from the general membership of the Natural Resources Defense Council.

Mothers and Others for Pesticide Limits

Meryl Streep, Chair
Wendy Gordon Rockefeller, Vice-Chair

National Committee

Eddie Albert, Actor and Conservationist
T. Berry Brazelton, M.D.
Meredith Brokaw, President, Penny Whistle Toys, Inc.
Robert Coles, M.D.
Joan Ganz Cooney, Chairman and CEO, Children's Television Workshop*
Sarah Jane Eads, Eads Ranch
Marian Wright Edelman, President, Children's Defense Fund*
Julie Nixon Eisenhower, Writer
Peter D. Hart, Chairman and CEO, Peter D. Hart Research Associates, Inc.
Teresa Heinz, Chair, National Council for Families and Television*
Terry Holsapple, Earthborne Farm
Michael F. Jacobson, Ph.D., Executive Director, Center for Science in the
    Public Interest
Rhoda Karpatkin, Executive Director, Consumers Union*
Penelope Leach, Ph.D., Fellow, B.P.R.S.
Eleanor Bingham Miller, Producer
Herbert Needleman, M.D., Committee on Environmental Hazards,
    American Academy of Pediatrics
Anna Quindlen, Writer
Richard Rivlin, M.D.
Maria Rodale, Promotion Coordinator, Rodale Press
Mrs. George Rublee, National Affairs and Legislation Committee,
    Garden Club of America
Courtney Kennedy Ruhe
Orville Schell, Writer
Iris Shannon, M.D., President, American Public Health Association
Richard and Sharon Thompson, The Thompson Farm
Manya S. Ungar, President, National PTA
Alice Waters, Owner and Chef, Chez Panisse

*Affiliation for identification purposes only

Natural Resources Defense Council

John H. Adams, Executive Director

Contents

For Our Kids' Sake

Foreword

We are endangering the future health of our children. The information in this book about the hazards of pesticides in children's food is bound to frighten parents of small children. And in fact, all of us *should* be worried about how our misuse of the environment, including our overuse of pesticides in growing our food, will affect our children's future.

But we must not let our fear make us feel powerless and anxious. Parents may feel helpless, and even overwhelmed, as I do, by the guilt they feel in not being able to protect their children more adequately. Grandparents like me are likely to feel angry and incompetent because we've allowed our environment and food supply to become so contaminated. But we can do our children even more harm with our anxiety and indecision. Instead, we need to look for short-term and long-term action. With action, we can feel less anxious and more effective. Our children need to see that we can and will have an effect on our world. Ultimately, our fear and guilt can be used as a positive motivating force.

So what can we do? I can't recommend that parents stop feeding children important foods like fruit and vegetables. Using vitamin substitutes for food certainly won't solve the problem. Instead, I'm very encouraged by the Solutions section in this book, which gives parents practical suggestions for reducing the pesticide residues in the food they feed their children. I recommend that you take those steps immediately, while continuing to serve your family a well-rounded diet of fruit, vegetables, and protein. As a consumer, you can search for and demand food which is grown without pesticides. We need to look for ways to bring down the prices of such food; not only the rich deserve this protection.

In addition, parents have to start working together on long-term solutions. That's why I, and other Americans who care passionately about our children's future, have joined the Mothers and Others for Pesticide Limits campaign. This campaign is a project of the Natural Resources Defense Council, which has a solid record of research and advocacy on behalf of the environment and public health.

The reforms recommended in this book by Mothers and Others are important ones. They are bound to be contested by major interest groups, including pesticide manufacturers and the food industry. We will be threatening their economy, so we can expect a difficult battle.

But a strong public outcry will produce results. We've already seen results in similar battles to curb apple coloring techniques and to reduce lead exposure where small children play. Unfortunately, such changes reach the poor, most vulnerable families last. But we must start now to work for reforms. Our children's future health is too precious for us to wait.

Let's all turn our anger and our helplessness into action to protect our children and to institute practices and controls that will clean up our children's food.

—T. Berry Brazelton, M.D.

Oats, peas, beans, and barley grow
Oats, peas, beans, and barley grow
Can you or I or anyone know
How oats, peas, beans, and barley grow?

—*Children's nursery rhyme*

Introduction

As caring parents, we are constantly vigilant about our children's welfare. We keep their safety in mind when we choose toys for them; we protect them from accidents by "childproofing" our homes; we have them vaccinated against diseases and give them proper medical care. And we feed them nutritious food.

That is, we *try* to choose healthy food. The truth is that lurking in the fresh, tasty-looking vegetables and fruit that we feed our children—the very same produce that contains nutrients essential for their growth—is the invisible danger of pesticides. A major new report by the Natural Resources Defense Council (NRDC) reveals just how serious that invisible danger is. According to the report, preschool children receive even greater exposure than adults to toxic pesticides in their food, precisely at the time in their lives when they may be most vulnerable to the harmful effects of those poisons.

This exposure translates into alarming statistics: NRDC estimates that more than 5,000 of today's preschoolers may get cancer sometime in their lives solely because of the pesticides in food they will have eaten by the time they are just six years old. And that figure almost certainly understates the problem, since it's based on an examination of only *eight* of the *more than 300* pesticides that the government itself has allowed for use on food. What's more, the NRDC report warns that at least three million of the country's preschool children may currently be exposed through their food to "neurotoxic" pesticides, which can harm the nervous system, at levels above what the federal government considers safe.

The sad irony is that most of this routine over-exposure to pesticides is viewed by our own government as entirely *legal*. Children are clearly not being adequately protected, and the blame lies squarely with our government, which has failed to enforce the laws designed to keep the food supply safe from harmful pesticides. Government has failed us in another way as well: Agricultural methods in the U.S. have become dangerously reliant on pesticides, and the government has not provided farmers with incentives to switch to safer, effective techniques that could dramatically reduce, or even eliminate, pesticide use.

It's up to us—parents, consumers, and citizens—to insist that our children be protected. *For Our Kids' Sake* is an action-oriented book written to give you the information and practical tools you need to start working immediately to protect your children against pesticides. In the short-term, this book details some simple measures that

Opposite page: Insecticide spraying by plane over a field of green peas.

you can start taking today, right in your kitchen, to reduce the amounts of pesticides in your children's food. But what's really needed are long-term solutions, including basic reforms in the way pesticides are regulated in this country; changes in the marketplace to ensure that healthy, pesticide-free food is widely available; and major reductions in the use of pesticides in producing the country's food supply. The following pages describe these solutions and tell you how *you* can work to make these changes happen.

How to use this book

For Our Kids' Sake is divided into three main sections: **The Problem**, **Solutions**, and **Tools**. In addition, the book features a center tear-out **Action** section.

Section 1, **The Problem**, describes the problems with pesticides and why they are particularly hazardous in children's food, and details the many ways in which government regulation has failed to protect children from over-exposure to pesticides. The **Solutions** section outlines the most important steps that the government must take to start protecting children, as well as what individuals should do—at home, in the marketplace, and as citizens—to protect children and to press for necessary reforms. The final section, **Tools**, is full of more detailed information that you can use in working on this issue, including further information on the most common and hazardous pesticides, practical steps for getting organic produce into your supermarket and for working with farmers to improve food safety, sources of pesticide-free produce, and tips on lobbying and organizing. The **Tools** section also includes a glossary explaining technical terms; the first time such terms appear in the book, they are highlighted with bold type.

You may want to start out by reading the sections on **The Problem** and **Solutions**—they lay out the issue clearly and understandably. Or you may choose to turn directly to the center **Action** section, to "7 ways to protect your family against pesticides in food," which describes tips on buying and preparing food to reduce your children's exposure to pesticides. We suggest that you attach this sheet to your refrigerator or another prominent place in your kitchen and start following the recommendations immediately. Note, though, that even careful washing, peeling, and cooking of produce may not remove all the pesticide residues; the charts beginning on page 38 of the **Tools** section provide more detailed information on which techniques work and don't work for specific produce and pesticides.

Next, we urge you to sign the mail-in lobbying cards in the

Action section, and send them to Congress and the government agencies to press for crucial reforms in pesticide use and regulation. Also included in that section is a form for you to notify NRDC that you have mailed the cards; this will help us in our own efforts. You can use the same form to join our national campaign, Mothers and Others for Pesticide Limits. With your support, you'll be joining your voice with many others in pressuring government for reforms. You'll also receive a quarterly newsletter as well as useful updates and timely alerts on what we can do about pesticides—*for our kids' sake.*

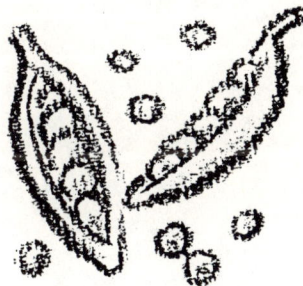

Section 1.
The Problem

Pesticides are by definition *poisons*, used to kill or control pests. There are **insecticides** to control insects, **fungicides** to control fungi such as mold and mildew, **herbicides** to control weeds, and **rodenticides** to control rodents. Some of these pesticides help increase crop yields or prolong the storage life of produce; others are simply used in the quest for perfect-looking fruit and vegetables. For instance, in growing citrus, the most widely applied pesticides are used to control a pest that some experts believe causes only cosmetic damage.

Unfortunately, pesticides do more than kill unwanted pests. They also destroy harmless insects that could otherwise serve as natural pest controllers. And more important, pesticides are potentially harmful to humans. They have been known to cause **cancer**, nerve disorders, birth defects, and genetic defects.

Many of the chemical pesticides used today were first developed for chemical warfare during World War II. In the post-war years, pesticides were heralded as a miracle of sorts, and were used more and more frequently, with little knowledge of and regard for health or environmental hazards. By the time the dangers of once-popular pesticides like DDT were discovered, irreversible damage had already been done. DDT and similar **organochlorine** pesticides were found to **persist** almost indefinitely in the environment, moving up through the food chain from plants to animals and humans. Today, even though DDT use has been banned in the U.S. since 1972, nearly every American still has traces of DDT in his or her body.

And yet the use of pesticides has increased dramatically: The amount of pesticides used in the U.S. has multiplied ten-fold since the 1940s, and has doubled in the past two decades alone. More than two and a half billion pounds of pesticides are now used each year—on agricultural crops, in forests, on ponds and lakes, in city parks, on lawns, and in homes. In 1985, farmers, homeowners, and commercial pest control companies spent a total of $6.6 billion on pesticides. As a result of their widespread use, there is pesticide contamination of our air, our drinking water supply, *and* our food.

The increasing use of these chemicals and their pervasive pres-

Reminder: There is a glossary on pages 86 and 87. When terms in the glossary first appear in this book, they are highlighted in bold type.

Opposite page: Helicopter spraying a fungicide over an orange grove.

ence in the environment and water and food supply is taking its toll. As a result of normal agricultural use, 46 pesticides (including some **carcinogens**) have been found contaminating groundwater in at least 26 states. DDT has been detected in animals from areas like the Antarctic, where it was never sprayed. Meanwhile, some insects have become major pests only after chemicals have killed their natural predators. Other insects and pests have developed immunities to pesticides; since World War II, the number of insect species known to be resistant to an insecticide has grown to more than 440. There are now more than 20 species of insects that no known pesticide will kill. In 1984 alone, pesticide resistance cost farmers $150 million in crop damage and increased use of chemicals.

There are also high *health* costs of increased pesticide use. Studies show higher cancer risks among farmers who are routinely exposed to pesticides. In California, pesticides are the major single cause of occupational illness. And it isn't just the *users* of pesticides who are at risk. A 1987 study for the National Cancer Institute indicated that children living in homes where household and garden pesticides are used are seven times as likely to develop childhood leukemia as other children. But the real extent of health damage from pesticides isn't even known, since the majority of pesticides used today haven't been properly tested for health hazards.

Our children at risk:
Pesticides in children's food

Similarly, because of inadequate testing, no one knows for sure the extent of pesticide contamination of food. In 1987, the federal Food and Drug Administration (FDA), which is responsible for monitoring food for pesticides, found pesticide **residues** in 50 percent of the fruits, and 41 percent of the vegetables it sampled. But actual contamination is probably much worse, since the agency samples less than one percent of the food supply, and since its commonly-used laboratory methods can detect less than *one-half* of the pesticides that are likely to leave residues on food. In fact, the FDA's routine sampling can't detect 40 percent of the pesticides that the agency itself considers to be health hazards.

In early 1989, the Natural Resources Defense Council completed the most comprehensive analysis ever conducted—either privately or by government—of children's food consumption and pesticide residues in children's food. The findings of the two-year NRDC study, "Intolerable Risk: Pesticides in Our Children's Food," are alarming; among other problems, the report found that:

Fact:
Because of agricultural use, at least 46 pesticides—some of them cancer-causing—have been detected in groundwater in 26 states.

• The average child receives four times more exposure than an adult to eight widely-used carcinogenic, or cancer-causing, pesticides in food. Because of their exposure to these pesticides alone, as many as 6,200 children may develop cancer sometime in their lives. And these eight pesticides are just a fraction of the 66 pesticides that the EPA has identified as potentially carcinogenic that might be found in a child's diet.

The greatest source of cancer risk identified by NRDC comes from apples, apple products, and other foods, such as peanut butter and processed cherries, that may be contaminated with daminozide and its **metabolite**, UDMH (what daminozide breaks down into during processing). In fact, average exposure to UDMH may cause one cancer case for every 4,200 children exposed by the age of six alone—which is *240 times* the cancer risk that the EPA considers "acceptable" following a *full lifetime* of exposure. Among other carcinogens posing risks to children are the fungicide mancozeb and its metabolite ETU, found in foods like tomato products and apple juice; and captan, a fungicide used on crops such as strawberries.

• At least 17 percent of the country's 18 million one- to five-year-olds are being exposed to **neurotoxic organophosphate** pesticides—which are designed to poison insects' nervous systems and can cause nervous system damage or impair the behavioral system in humans—at levels above what the federal government considers safe. Children's exposure to organophosphates comes through tomatoes and tomato products, green beans, orange juice, and cucumbers, among others.

The problem of pesticides in food is particularly serious for children for several reasons. For one thing, children consume proportionally more fruits and vegetables—and therefore more pesticides—than adults do. Children don't actually eat more food than their parents, in absolute terms, but relative to their body-weight (which is the more relevant measure for determining pesticide exposure), they consume more of most foods than adults. Produce makes up about one third of the average child's diet, and the typical preschooler's diet is dominated by fruits, which are the foods *most* likely to be contaminated by pesticide residues. In fact, relative to body-weight, the average preschool child eats *six times* the amount of fruit and fruit products that an average adult woman eats—including 18 times the amount of apple juice. (Young infants consume even more apple juice—on average, *31 times* more than an adult woman consumes.) The charts on pages 20 and 21 show the relative food consumption by children and women for various foods, and the differences in exposure to a number of pesticides.

Food Intake
Children and Women

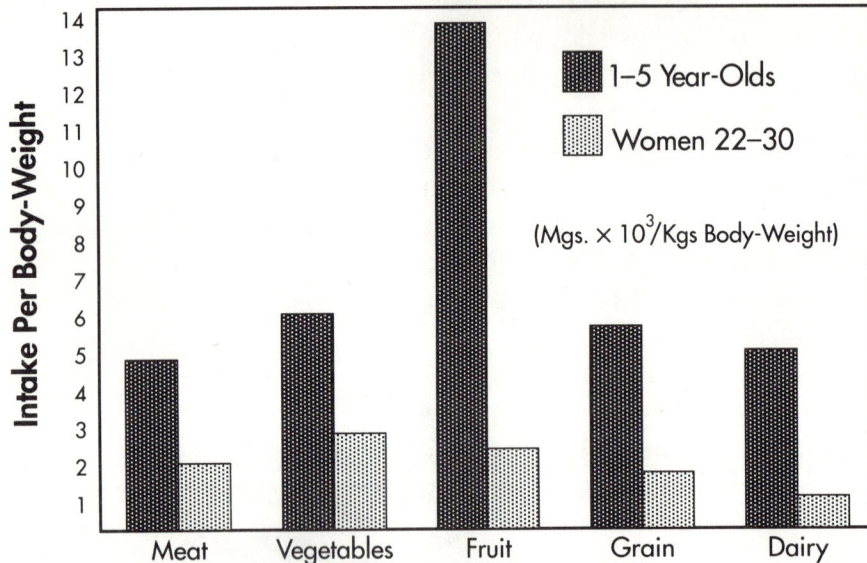

Intake Per Body-Weight

14
13
12
11
10
9
8
7
6
5
4
3
2
1

■ 1–5 Year-Olds

▨ Women 22–30

(Mgs. $\times 10^{3}$/Kgs Body-Weight)

Meat   Vegetables   Fruit   Grain   Dairy

Children's greater exposure to pesticides comes precisely at a time in their lives when they may be especially vulnerable to pesticides' toxic effects. Children are probably more susceptible to neurotoxic pesticides for several reasons, including the fact that their nervous systems are still developing. In addition, studies have shown that immaturities in the digestive systems of the young mean that they absorb toxic chemicals more readily than adults do, and the organs of the young may be more susceptible to the toxic action of various chemical compounds.

What's more, children are probably more vulnerable than adults to carcinogens. There are two reasons for this. First, cells divide more rapidly in infancy and early childhood, which increases the probability that cell mutations can be passed to subsequent generations of cells and can therefore start the development of cancer. In addition, earlier exposure to carcinogens simply means there's more time for cancer to develop. In fact, more than half of the lifetime cancer risk that an individual faces from pesticide residues in fruits may be from exposures in just the first six years of life.

Government's failure to protect children: Problems with pesticide regulation

The blame for this unacceptable threat to our children lies with government neglect and with the chemical industry's inordinate influence over pesticide regulation. The Environmental Protection Agency is responsible by law for establishing safe limits for pesticides in food, while the Food and Drug Administration is charged with monitoring food to detect pesticide residues and with enforcing the limits established by the EPA. Their combined failure to protect the general public from pesticides has been criticized in study after study by congressional committees, the government General Accounting Office, and the National Academy of Sciences. But the NRDC report is the first to point out in a comprehensive way the particularly serious failure of the government to protect *children* from pesticides in food. The report cites many serious loopholes in the current regulation of pesticides, including:

Regulatory loopholes at the EPA

• **The EPA did not take children's eating patterns into account when it set virtually all the current legal limits for pesticides in food.** Instead, almost all of these limits were based exclusively on estimates of *adults'* food consumption levels— and largely outdated estimates,

Differences in Relative Exposure
(Children vs. Women)

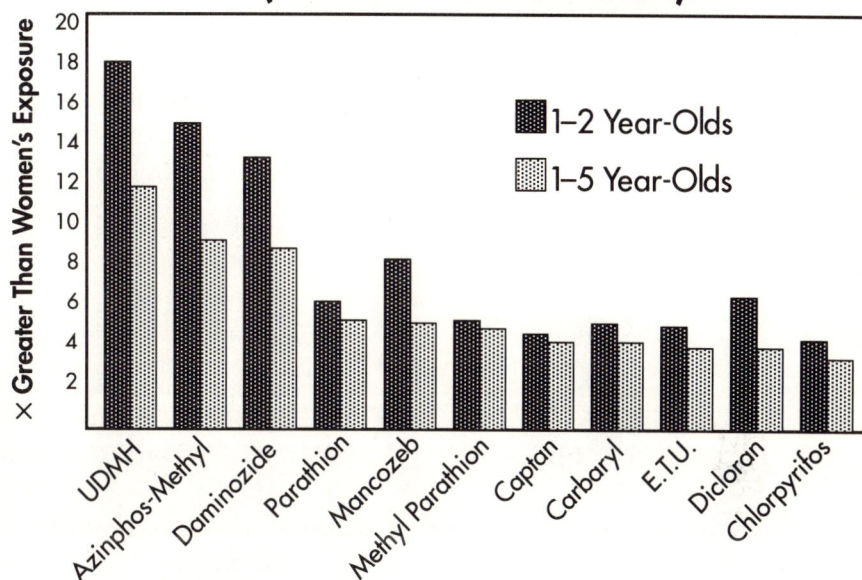

at that. In setting almost all the current legal limits (known as **tolerances**) for pesticide residues in food, EPA has relied on estimates of adults' food consumption called "Food Factors," which were developed in the late 1960s. Since consumption of fresh produce has increased significantly in recent decades, Food Factors underestimate even adult produce consumption today. (For example, the government assumes that an adult consumes no more than 7½ ounces *per year* of avocados, artichokes, eggplants, melons, or tangerines.) And Food Factors seriously underestimate *children's* food consumption—particularly children's consumption of fruit. Food Factors underestimate preschoolers' consumption of cranberries by 14-fold, grapes by six-fold, apples and oranges by about five-fold, apricots by almost four-fold, and strawberries by almost three-fold. They also underestimate children's consumption of most vegetables— including carrots by two-fold, and broccoli by more than two-fold.

• **Because the EPA underestimates children's consumption of most foods, pesticide exposure levels in children's food that our government claims are legal can greatly exceed safe levels.** By law, the EPA is required to set limits for pesticides in food at levels that are safe. But the EPA's formula for setting most of these exposure limits is based on assumptions that drastically underestimate children's food consumption. As a result, many of the maximum "acceptable" amounts of pesticide residues in food are simply not safe for children to eat. For instance, a child's exposure to the pesticide demeton at the limit set by the EPA can be 400 times higher than the level that the EPA has determined to be safe for adults. For disulfoton, exposure for children can be 180 times the "safe" level; and for diazinon, 160 times higher. In fact, if all of today's preschoolers were exposed at the maximum levels permitted by the EPA to just three fungicides—captan, folpet, and mancozeb—more than 20,000 children could eventually develop cancer during their lives as a result of that exposure during their preschool years alone.

• **Adequate testing hasn't been done on the health effects of many pesticides.** According to the Federal Insecticide, Fungicide and Rodenticide Act (FIFRA), before a pesticide can be licensed, the EPA has to determine that it won't cause "unreasonable adverse effects" on human health or the environment. The EPA currently requires pesticide manufacturers to submit as many as 150 tests on environmental and health effects of pesticides. But *most* pesticides used today were registered either before modern testing requirements were in place, or before EPA even existed. As a result, the vast majority of the 600 **active ingredients** used today in 50,000 pesticide products haven't been adequately tested. In 1972, Congress required the EPA to "reregister" pesticides, but the process has been extremely slow; the General Accounting Office has estimated that 390 of the 400 pesticides approved for food use are older pesticides that are only now being reassessed. FIFRA amendments passed by Congress in 1988 address the problem of the slow reassessment process by requiring that all such testing be completed within the next nine years. However, while they're being tested, the pesticides are still in use.

Another problem with EPA testing requirements is that the Agency doesn't require important testing for "neurotoxic" effects. In testing for the toxic effects of pesticides that act on the nervous system, the EPA requires just one test, to see if a pesticide causes a particular paralytic reaction following only a short exposure at very high levels. The EPA doesn't require testing for the effects of long-term, low-level exposures to pesticides on the developing nervous system, and on such things as learning ability, emotions, sight, hearing, and memory.

• **Even pesticides that are known to be hazardous are allowed to stay on the market while under special EPA review.** When the EPA receives new data showing that a pesticide poses a hazard to the environment or human health, it initiates a **Special Review** process to determine if use of the pesticide should be restricted. Ironically, though, pesticides are allowed to stay on the market while under EPA Special Review, and their uses can even be expanded. The review is supposed to determine quickly if a pesticide should be restricted, but instead it generally takes from two to six years to complete and some reviews have taken 10 years. The pesticide captan was in Special Review for nine years, for instance, while the EBDCs have been in Special Review since 1977. Meanwhile, children continue to be exposed to the pesticides' hazards; during a six-year Special Review, 22 million children can pass through their preschool years. Of the pesticides reviewed by NRDC, those which are currently under Special Review together pose up to 98 percent of the carcinogenic risk to the average preschool child.

• **The EPA allows significant levels of cancer-causing pesticides in food.** In part, this is because carcinogenic pesticides remain on the market while under Special Review. But the EPA's interpretation of pesticide laws is also to blame. Under the "Delaney Clause" of the Federal Food, Drug and Cosmetic Act, an additive can't be allowed in food if it has been found to induce cancer in humans or animals. Pesticide residues in raw produce, feed, and some processed foods are exempt from that standard, however; instead, the law requires that individual legal limits be set for pesticides in those cases, and that the limits be set to protect public health while considering the need for an "adequate, wholesome, and economical food supply." The EPA has interpreted this to mean that it must weigh cancer risk against economic benefits, an approach that has been criticized in a number of studies—including studies by Congress and by the National Academy of Sciences—and challenged in a pending court action by NRDC, Public Citizen, and Ralph Nader.

• **The EPA's methods for setting legal limits for pesticide residues in food don't adequately protect children.** In determining the cancer risks of pesticides, the EPA relies on animal studies with exposures starting at a point that is comparable, in human life, to missing the critical preschool years. In addition, when setting the

Fact:
Strawberries and peaches have been found to contain high levels of pesticide residues, probably because heavy spraying is needed to preserve the fruit's appearance.

safe levels of pesticide exposure, the EPA uses safety margins that may not provide enough protection for the most sensitive members of the population—children. Still another problem is that in setting exposure limits, the EPA doesn't consider the health effects of **inert ingredients** (which are labelled "inert" because they have no pest-killing action, not because they are *harmless*). The trouble is that while some inert ingredients *are* harmless, many of them—such as the known carcinogens methylene chloride and benzene—pose serious health risks.

Enforcement loopholes at the FDA

• **In enforcing legal restrictions on pesticide residues, the FDA food sampling program is extremely limited.** The agency itself collects and analyzes about 12,500 samples of produce, milk, eggs, fish, processed foods, and animal feed each year, to determine whether pesticide residue limits are exceeded. This amounts to slightly more than one percent of imported produce, less than one percent of domestically grown produce, and proportionally even less of processed products that children consume in large quantities. While there are state programs for sampling food for pesticides, information sharing between the states and the FDA is inconsistent, and the state sampling programs are often limited both in the numbers and levels of pesticides that they can detect.

• **The tests used by the FDA for analyzing pesticide residue levels are too limited.** The FDA has developed five tests for detecting various pesticide residues. But collectively these tests can only detect about 40 percent of the pesticides that may leave residues in food, and in fact, the FDA normally uses only one or two of the tests for each sample. Typical testing is limited to only about 100 of the approximately 500 pesticides that may leave residues in food. Of the pesticides that the FDA itself classifies as having a moderate to high health hazard, 40 percent cannot be detected by any of the five FDA tests. In addition, the FDA sampling cannot detect many metabolites of pesticides, which poses a particular problem in sampling processed food. Although processing often reduces pesticide residues in a food, it can increase the concentration of certain pesticide residues or speed up the formation of toxic pesticide metabolites.

• **The FDA monitoring program is plagued with delays and doesn't serve as an effective deterrent.** According to the General Accounting Office (GAO), FDA laboratories can take an average of 28 days to complete and process sample analysis. By this time,

Fact:
The EPA has announced that at least 66 of the 300 pesticides used on food are potentially carcinogenic, but has not announced any action on restricting how these pesticides are used.

effective action is probably impossible, since the contaminated food has most likely already been sold and eaten. In fact, the GAO found that in more than half of the cases where the FDA found violations, the contaminated food was not recovered. What's more, the FDA penalized growers in only a few cases where it found violations.

Section 2. Solutions

Our children are at risk because of pesticide residues in the very food that we feed them to help them to grow up healthy. It doesn't have to be this way. But solving the problem of pesticides in children's food is going to require enormous public pressure on the government and the country's agricultural interests.

Public pressure *can* bring about needed changes in food safety. One example of the potential for success—but also of the critical need for government action—is the case of daminozide (trade name Alar). In 1986, environmental and consumer groups called attention to the widespread spraying of apples with Alar, which is used to deepen the red color of apples and to regulate growth so that they don't drop from trees before harvest, but which breaks down into the carcinogen UDMH during processing. The EPA has yet to ban the use of Alar, but the public outcry against Alar has been so strong that many growers have announced that they have stopped using it, several supermarket chains now say that they refuse to sell apples treated with Alar, and some food manufacturers have announced that they no longer use Alar-treated apples in their products. Unfortunately, even with these assurances there is no *guarantee* that food is Alar-free until the substance is banned.

Public pressure has led to other changes as well. In response to concern about pesticides in baby food, the H.J. Heinz Company has written to growers that it will not use food treated with 13 chemicals under EPA's Special Review in manufacturing its baby food. And because of consumer demand for safer food, some supermarket chains are now stocking organic produce. Raley's, a chain of 50 supermarkets in northern California and Nevada, features organically-grown fruit and vegetables. All Lucky Stores in northern California now routinely offer organic produce, and Safeway is test-marketing a selection of organic produce in 20 of its San Francisco area stores. Valu Food Supermarkets, the largest supermarket chain in Baltimore,

Tip:
Imported produce may contain more pesticides than domestically-grown produce, including pesticides that are banned for use in the U.S. FDA sampling of tomatoes found pesticide residues in 23 percent of domestic tomatoes, and in *70* percent of imported tomatoes.

began selling organically-grown produce in October 1988, and D'Agostino Supermarkets began offering organic produce in its 24 New York-area stores in early 1989. More than 100 Stop 'N Shop stores in Massachusetts and Connecticut now offer organic fruit and vegetables, and Krogers, a national supermarket chain, is stocking organic produce in its Ann Arbor, Michigan stores.

Other major supermarket chains have indicated that they will stock organic produce if the demand exists. So the ball is in our court; now it's up to us to *create* that demand for safe food.

What follows is an outline of the most important reforms needed to reduce pesticides in food, and some concrete steps that individuals can take to press for these reforms and to protect their families in the meantime. For a more complete description of the legislative reforms recommended by NRDC, see page 69 in the **Tools** section.

What government must do: Closing the loopholes in pesticide regulation and enforcement

Americans have the right to expect that *legal* food is *safe* food. Congress has to act immediately to protect that right—and to protect children—by overhauling the way the EPA regulates pesticides in food and the way the FDA enforces pesticide limits. Here's what needs to be done:

• The EPA should be required to quickly set new legal limits on exposure for many pesticides, to ensure that children and the general public are protected.

• In setting legal limits on pesticide exposure, the EPA must take children's actual eating habits into account, and establish limits that will really protect children. The EPA should be required to use the most current estimates of food consumption, and to revise old pesticide exposure limits and set new limits based on these more accurate estimates.

• In determining safe and legal pesticide exposure levels, the EPA should be required to take into account *cumulative* exposure to pesticides—including exposure to the variety of pesticides found in food, as well as pesticides in drinking water and in homes.

• The EPA should be required to take pesticides off the market quickly when they are found to pose serious risks. The Agency should also be able to ban the use of a pesticide if the manufacturer doesn't provide required data on it within the EPA's deadlines. And the EPA should prohibit the use of dangerous inert ingredients in pesticides.

• The EPA should require more complete testing of all pesticides for "neurotoxic" effects, such as effects on motor coordination, memory, learning, and intelligence.

• The FDA should improve its methods for sampling food and detecting pesticide residues, and the FDA's budget for pesticide monitoring should be increased. In addition, pesticide manufacturers should be required to develop practical and effective methods for the FDA to use to detect pesticide residues in food.

• The FDA must cut delays in analyzing food samples, and develop a system for tracking food found to contain **illegal residues**. The FDA should also be authorized to halt the shipment of food that is suspected of being illegally contaminated, and to impose civil penalties on violators.

• The FDA should require food retailers to label fresh produce with the country of origin, and to provide consumers with information on the pesticides used in growing, transporting, or storing any produce being sold.

What farmers can do (with government's help): Reducing pesticide use

In the long-term, the best way to eliminate hazardous exposure to pesticide residues in food will be to reduce their use in growing crops. Faced with the dangers and high costs of their increased reliance on pesticides, many farmers would probably be willing to modify their growing methods if it weren't for the financial risks involved initially. Several basic reforms in farming policies could give farmers the incentives and financial assistance they need to make the switch to safer farming techniques:

• **The federal government should provide growers with credit assistance, crop insurance, and other financial protection during the transition from chemical-intensive agriculture to innovative, safer farming techniques.**

Two such techniques are organic farming and **integrated pest management.** Organic farming generally means that growers use no chemical pesticides. "Integrated pest management" (IPM) generally means that growers don't routinely spray with pesticides, but instead monitor crops carefully for pests and use chemical pesticides only when necessary—for instance, when pest populations threaten to damage a crop. Both organic and IPM growers rely on a variety of alternatives to pesticides, including planting crop varieties that resist pest damage, and using pests' natural enemies in pest control. IPM techniques are already being used successfully on specific crops, with cuts in pesticide use ranging from 25 percent to more than 80 percent. A 1986 report by the congressional Office of Technology Assessment estimated that IPM techniques could cut pesticide use by as much as one-half.

These alternative farming techniques that eliminate or reduce the need for pesticides may actually be more profitable than conventional farming in the long run. But farmers switching from conventional methods to these low-chemical methods could suffer short-term losses while they learn and perfect the new techniques. Guaranteed government assistance could help ease the transition.

• **Pesticide products should be taxed to fund research and demonstration projects in alternative, low-chemical farming.**

In 1988, less than $4 million of the Department of Agriculture's $1 billion annual research budget went toward the USDA's new research program in alternative agricultural techniques. A tax of just one percent on pesticide products could raise tens of millions of dollars in revenue for research and demonstration projects on techniques for reducing reliance on pesticides.

• **To protect consumers, Congress should establish a uniform national definition of "integrated pest management" and "organic" farming.**

Right now, some states have certification programs that define techniques like organic farming, but the definitions aren't necessarily consistent from state to state (see page 60 in the **Tools** section). As a result, consumers can't know for sure that food labeled "organically-grown" really was grown without pesticides. A nationally consistent definition will help provide consumers with this assurance.

• **Farmers shouldn't be penalized for using crop rotation and other techniques that can reduce pesticide use.**

Crop rotation can help prevent pest problems, and therefore reduce the need for pesticides. But under current federal commodity programs, farmers who qualify for support payments actually risk losing their eligibility if they adopt crop rotation. Congress should reverse this policy, and should also ensure that crop insurance policies don't discriminate against farmers who use organic or integrated pest management techniques.

• **Congress should prohibit agricultural "supply-control" programs from requiring cosmetically perfect produce.**

Currently, marketing and grading standards developed by the agricultural industry require commercial produce to meet very exacting criteria for color, size, shape, and texture. But produce meeting these criteria may not necessarily be the most nutritious or best-tasting produce—*or* the safest. In fact, growing produce that *looks* perfect may require excessive pesticide use.

Tip:
Think twice before buying perfect-looking produce. It may look that way because of heavy pesticide use.

What *you* can do

None of the above reforms will come about without sufficient public pressure, both on Congress and in the marketplace. One individual working alone can't have much impact on a problem as large as this one, but many people working individually and together on this issue *can* make a difference—by writing to elected officials and government agencies to urge reforms, writing letters to newspaper editors to call attention to the problem, and forming local groups to talk to supermarket managers about stocking organic produce. Here's how you can start working today to protect your children against pesticides in food:

• **As a parent,** you can start right now, at home, by following a few simple steps in choosing and preparing food to reduce your family's exposure to pesticide residues. The tear-out sheet, "7 ways to protect your family from pesticides in food," in the center **Action** section of this book, explains what to do.

• **As a consumer**, you can exercise your power in the marketplace and work to get your supermarket to stock pesticide-free produce. The expansion of alternative farming methods that reduce or eliminate pesticide use, such as organic farming or integrated pest management, will depend in part on consumers deciding that pesticide-contaminated food is unacceptable, and acting on that decision by buying only pesticide-free food. The **Tools** section includes information on how to get your store to stock organically-grown food, as well as direct sources for buying organic food.

• **As a citizen,** you can lobby Congress, the EPA, the FDA, and the Department of Agriculture (USDA) for crucial reforms in how pesticides are regulated, and for changes in the country's farm policies to encourage growers to reduce the use of pesticides on food crops. Included in the **Action** section are cards for you to sign and mail to Congress, the EPA, and the FDA. If you have time to write longer, personalized letters to urge these reforms, or if you'd like to encourage neighbors and friends to write, you'll find some tips on letter-writing on page 67 in the **Tools** section.

• You can join NRDC's **Mothers and Others for Pesticide Limits** campaign. We're working nationally to call attention to the problem of pesticides in children's food, to press for reforms in pesticide regulations and enforcement, and to ensure that safe produce is widely available around the country. There's a form to use in joining our campaign in the **Action** section. As a supporter, you'll be kept informed through our quarterly newsletter and other alerts and updates about environmental and food safety issues affecting your children. And you'll be joining with many others in sending decision-makers the powerful, unified message that parents demand safe food.

Section 3. Tools

Pesticides that are the greatest sources of exposure to children*

Chemical	Potential health effects
Mancozeb	"Probable human carcinogen"; also contains and breaks down to ETU (see below); mutagen; causes birth defects in experimental animals; effects on the kidney, thyroid and prostate glands.
Daminozide	"Probable human carcinogen"; also contains and breaks down to UDMH (see below); causes multiple tumors at multiple organ sites (lung, liver, kidney, reproductive and vascular systems) in animals.
Carbaryl	Mutagen; causes kidney effects in animals.
Captan	"Probable human carcinogen"; mutagen; causes reproductive effects in animals; possible teratogen.
Dicloran	Liver effects in animals.
Folpet	"Probable human carcinogen"; causes decreased weight gain and blood constituent level changes in animals.

Adapted with permission from "Intolerable Risk: Pesticides in Our Children's Food," by B. Sewell, R. Whyatt, J. Hathaway, and L. Mott, NRDC, 1989

*Of the 23 pesticides and metabolites evaluated by NRDC in children's consumption of 27 types of fruits and vegetables

Chemical	Potential health effects
ETU	"Probable human carcinogen"; affects thyroid gland and causes other hormonal effects in animals; causes birth defects in animals.
Chlorothalonil	"Probable human carcinogen"; mutagen; causes reproductive and kidney effects in animals.
Parathion	"Possible human carcinogen"; mutagen; extremely toxic; causes nervous system effects (interferes with nerve transmission function by inhibiting enzyme ChE, degenerative effect on nerve fibers) and eye effects in animals.
Methamidophos	Extremely toxic (nervous system toxin—inhibits ChE); causes reproductive effects in animals.
UDMH	"Probable human carcinogen"; causes multiple tumors at multiple sites (lung, liver, pancreas, nasal tissue, vascular system) in animals; mutagen.
Chlorpyrifos	Causes nervous system effects (inhibits ChE); mutagen; eye and skin irritant.
Azinphos-methyl	Acutely toxic (nervous system toxin—inhibits ChE); severe eye and skin irritant; mutagen; causes cancer (liver tumors) in animals.
Dimethoate	Nervous system effects (inhibits ChE); mutagen; causes reproductive effects in animals; some evidence of carcinogenicity in animals.
Omethoate	Degradation product of dimethoate; more acutely toxic than dimethoate (nervous system toxin—inhibits ChE).
Methyl parathion	Acutely toxic (nervous system toxin—inhibits ChE); degenerative effects on nerve tissue; mutagen; causes birth defects and reproductive effects in animals; some evidence of carcinogenicity in animals; affects eyes (cataracts) in animal studies.
Permethrin	Eye irritant; liver effects in animals; causes lung and liver tumors in animals.

Fact:
During a six-year EPA Special Review of a pesticide, 22 million children can pass through their preschool years.

Section 3.
Tools

Pesticides that are
the greatest sources
of exposure to
children

Chemical	Potential health effects
Acephate	"Possible human carcinogen"; mutagen; mild eye irritant; causes nervous system effects (inhibits ChE); reproductive effects in animals.
Diazinon	Eye and skin irritant; causes nervous system effects (inhibits ChE).

Reducing pesticide residues in produce

Apples

Diphenylamine *DPA*	Residues remain primarily in the peel. DPA does not readily dissolve in water; therefore, plain water washing may not reduce residues, but peeling may help.
Captan* *Merpan, Orthocide*	Residues remain primarily on the produce surface. However, the metabolite THPI, a suspected carcinogen, may be systemic. Washing, cooking, or heat processing will reduce residues.
Endosulfan *Thiodan*	Residues remain primarily on the produce surface; however, endosulfan metabolites may be systemic. Peeling, cooking, or heat processing may reduce residues slightly. No information on removal with water.
Phosmet *Imidan*	Residues remain primarily on the produce surface. Washing or cooking will reduce residues.
Azinphos-methyl *Guthion*	Residues remain primarily on the produce surface. Washing, cooking, or heat processing will reduce residues.

Bananas

Diazinon *Spectrocide, Sarolex*	Residues remain primarily on the produce surface. No information on removal with water.
Thiabendazole *TBZ, Mertect*	Residues are primarily found in the peel. Peeling or washing will reduce residues.
Carbaryl *Sevin*	Residues remain primarily on the produce surface. Washing or peeling will reduce residues.

*Pesticides in bold type are especially hazardous.

Adapted with permission from *Pesticide Alert*, by L. Mott and K. Snyder, Sierra Club Books, 1987. (For information on ordering, see page 83.)

Note: These charts include the five pesticides detected most frequently in each fruit or vegetable by the FDA and California monitoring programs. Some pesticides that may be particularly hazardous to children, such as daminozide (and its metabolite UDMH) and mancozeb (and ETU), cannot be detected by the routine laboratory methods of these agencies, so they do not appear in these charts.

Broccoli

DCPA *Dacthal, Chlorthaldimethyl*	Some evidence that residues are systemic. For a preemergence herbicide, the relatively frequent findings of residues indicate a fair degree of persistence. No information on removal with water.
Methamidophos *Monitor*	Residues are systemic and probably cannot be removed with washing.
Dimethoate *Cygon, Rogon*	Residues are systemic. However, washing, peeling, cooking, or heat processing reduced residues in various studies.
Demeton *Systox*	Residues are systemic and probably cannot be removed with washing.
Parathion *Phoskil*	Residues remain primarily on the produce surface. Washing, peeling, cooking, or heat processing may reduce residues slightly.

Cabbage

Methamidophos *Monitor*	Residues are systemic and probably cannot be removed with washing. Residues remain primarily in the outer leaves, so stripping these tough layers in cabbage may reduce residues.
Dimethoate *Cygon, Rogon*	Residues are systemic. However, washing, peeling, cooking, or heat processing reduced residues in various studies.
Fenvalerate *Pydrin, Belmark*	Residues remain primarily on the produce surface. Fenvalerate does not readily dissolve in water; therefore, plain water washing may not reduce residues.
Permethrin *Ambush, Pounce*	Residues remain primarily on the produce surface. Washing with detergent will reduce residues; plain water may not.
BHC *HCH, 666, Hexachlor*	Residues remain primarily on the produce surface. No information on removal with water.

Tip:
Corn and cauliflower have been found in sampling to contain relatively fewer pesticide residues. The heavy leaves on cauliflower and the husks on corn may help protect the edible portions from pesticide exposure.

Cantaloupes

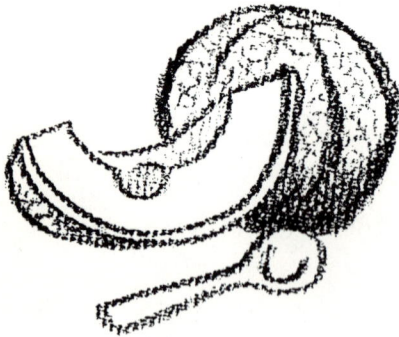

Methamidophhos *Monitor*	Residues are systemic and probably cannot be removed with washing.
Endosulfan *Thiodan*	Residues remain primarily on the produce surface; however, endosulfan metabolites may be systemic. Peeling may reduce residues slightly. No information on removal with water.
Chlorothalonil *Bravo*	Residues remain primarily on the produce surface; however, chlorothalonil metabolites may be systemic. Washing reduces residues.
Dimethoate *Cygon, Rogon*	Residues are systemic. However, washing and peeling reduced residues in various studies.
Methyl Parathion *Folidol M, Metacide*	Residues remain primarily on the produce surface, but some evidence that residues can be absorbed. No information on removal with water.

Carrots

DDT	Residues remain primarily on the produce surface, although residues may be absorbed into the peel. Washing and peeling reduce residues in root crops, but cooking won't reduce residues once they have been absorbed into the plant tissue.
Trifluralin *Treflan*	Residues are systemic and probably cannot be removed with washing, particularly in root crops. Peeling may reduce residues in carrots.
Parathion *Phoskil*	Residues remain primarily on the produce surface. Washing, peeling, cooking, or heat processing may reduce residues slightly.
Diazinon *Spectrocide, Sarolex*	Residues remain primarily on the produce surface. Cooking or heat processing may reduce residues. No information on removal with water.
Dieldrin	Residues remain primarily on the produce surface, although there is some evidence that residues are systemic in root crops. No information on removal with water. Peeling or cooking may reduce residues.

Note: DDT has been banned in the U.S. since 1972, but because of its persistence in the environment, DDT residues are still found in some produce, particularly root crops.

Cauliflower

Methamidophos
Monitor
Residues are systemic and probably cannot be removed with washing.

Dimethoate
Cygon, Rogon
Residues are systemic. However, washing, peeling, cooking, or heat processing reduced residues in various studies.

Chlorothalonil
Bravo
Residues remain primarily on the produce surface; however, chlorothalonil metabolites may be systemic. Washing or cooking reduces residues.

Diazinon
*Spectrocide,
Sarolex*
Residues remain primarily on the produce surface. Cooking or heat processing may reduce residues. No information on removal with water.

Endosulfan
Thiodan
Residues remain primarily on the produce surface; however, endosulfan metabolites may be systemic. Peeling, cooking, or heat processing may reduce residues slightly. No information on removal with water.

Celery

Dicloran
DCNA, Botran
Residues remain on surface following foliar treatment but are absorbed and translocated to edible tissue, following soil treatment. Incorporation of dicloran into wax formulations reduces the effectiveness of washing. Washing, peeling, cooking, or heat processing may reduce residues.

Chlorothalonil
Bravo
Residues remain primarily on the produce surface; however, chlorothalonil metabolites may be systemic. Washing or cooking reduces residues.

Endosulfan
Thiodan
Residues remain primarily on the produce surface; however, endosulfan metabolites may be systemic. Peeling, cooking, or heat processing may reduce residues slightly. No information on removal with water.

Acephate
Orthene
Residues are systemic and probably cannot be removed with washing. Cooking or canning may reduce residues.

Methamidophos
Monitor
Residues are systemic and probably cannot be removed with washing.

Cherries

Parathion
Phoskil

Residues remain primarily on the produce surface. Washing, peeling, cooking, or heat processing may reduce residues slightly.

Malathion
Cythion

Residues remain primarily on the produce surface, but may be absorbed into the peel. Washing with detergent reduces residues more than plain water. Peeling, cooking, or heat processing will reduce residues. Residues on dried fruits were higher than those on fresh fruits because of the concentration effect of dehydration.

Captan
*Merpan,
Orthocide*

Residues remain primarily on the produce surface. However, the metabolite THPI, a suspected carcinogen, may be systemic. Washing, cooking, or heat processing will reduce residues.

Dicloran
*DCNA,
Botran*

Residues remain on surface following foliar treatment. Incorporation of dicloran into wax formulations reduces the effectiveness of washing. Washing, peeling, cooking, or heat processing may reduce residues.

Diazinon
*Spectrocide,
Sarolex*

Residues remain primarily on the produce surface. Cooking or heat processing may reduce residues. No information on removal with water.

Corn

Sulfallate* *CDEC, Vegadex*	Sulfallate is not absorbed by foliage but is readily absorbed by roots and then moves through the plant. No information on removal with water.
Carbaryl *Sevin*	Residues remain primarily on the produce surface. Washing, peeling, and cooking will reduce residues.
Chlorpyrifos *Dursban*	Residues remain primarily on the produce surface. No information on removal with water.
Dieldrin	Residues remain primarily on the produce surface. No information on removal with water. Peeling or cooking may reduce residues.
Lindane *Agronexit, Lindafor, Gamma BHC*	Some evidence that residues are systemic. No information on removal with water.

Cucumbers

Methamidophos *Monitor*	Residues are systemic and probably cannot be removed with washing.
Endosulfan *Thiodan*	Residues remain primarily on the produce surface; however, endosulfan metabolites may be systemic. Peeling may reduce residues slightly. No information on removal with water.
Dieldrin	Residues remain primarily on the produce surface. No information on removal with water. Peeling may reduce residues.
Chlorpyrifos *Dursban*	Residues remain primarily on the produce surface. No information on removal with water.
Dimethoate *Cygon, Rogon*	Residues are systemic. However, washing and peeling reduced residues in various studies.

Fact:
DDT has been found in animals in the Antarctic and other areas where the pesticide was never sprayed.

*Pesticides in bold type are especially hazardous.

Grapefruit

Thiabendazole *TBZ, Mertect*	Residues are primarily found in the peel. Peeling or washing will reduce residues.
Ethio *Ethanox, Ethiol, Rhodocide*	Residues remain primarily on the produce surface. Washing or processing may reduce residues.
Methidathion *Supracide, Somonil*	Residues remain primarily in the peel of citrus fruit. No information on removal with water.
Chlorobenzilate *Acaraben*	Residues remain primarily on the citrus peel. Washing and processing of citrus fruit will reduce residues.
Carbaryl *Sevin*	Residues remain primarily on the produce surface. Washing or peeling will reduce residues.

Grapes

Captan *Merpan, Orthocide*	Residues remain primarily on the produce surface. However, the metabolite THPI, a suspected carcinogen, may be systemic. Washing, cooking, or heat processing will reduce residues. Wine made from captan-treated grapes had no residues in one study.
Dimethoate *Cygon, Rogon*	Residues are systemic. However, washing, peeling, cooking, or heat processing reduced residues in various studies.
Dicloran *DCNA, Botran*	Residues remain on surface following foliar treatment. Incorporation of dicloran into wax formulations reduces the effectiveness of washing. Washing, peeling, cooking, or heat processing may reduce residues.
Carbaryl *Sevin*	Residues remain primarily on the produce surface. Washing, peeling, or cooking will reduce residues.
Iprodione *Rovral*	Residues remain primarily on the produce surface. No information on removal with water.

Tip:
Trimming the leaves and tops off celery stalks may reduce pesticide residues.

7 ways to protect your family against pesticides in food

1. Wash all produce.

Washing fruits and vegetables in water may remove some surface pesticide residues; using a diluted solution of dishwashing soap and water may remove additional surface residues. Be sure to rinse well. Unfortunately, even careful washing may not remove all surface residues, and it can't remove pesticide residues that are contained inside the produce.

2. Peel produce when appropriate.

Peeling fruits and vegetables will completely remove surface pesticide residues. Remember, though, that peeling won't remove pesticide residues inside the produce, and that in some cases peeling may remove valuable nutrients.

3. Buy certified organically-grown fruits and vegetables.

You'll be accomplishing two things: You'll be feeding your family pesticide-free food, and you'll be sending supermarkets and growers the message that food laden with pesticides is unacceptable.

4. Buy domestically-grown produce in season.

Imported produce often contains more pesticides than domestically-grown produce, and may contain pesticides that are banned from use in the U.S. You can ask that your supermarket label produce for its country of origin, which should be easy to do since shipping containers usually identify the source of the produce.

5. Beware of perfect-looking produce.

Some pesticides are used simply to enhance the appearance of produce; flawless produce may signal that pesticides were used in growing it. A glossy surface may mean that the produce is waxed, and waxes can contain pesticides or may lock in surface pesticide residues.

6. Grow your own produce if possible.

You can avoid the use of pesticides or use only the ones that have been thoroughly tested and found not to cause adverse health or environmental effects.

7. Write to your elected officials.

In the long-term, the best way to protect your family against pesticides will be through reforms in the way pesticides are regulated, and reductions in pesticide use.

Mothers and Others
FOR PESTICIDE LIMITS

Boynton

A PROJECT OF THE
NATURAL RESOURCES DEFENSE COUNCIL

Sign, stamp, and mail these postcards today to urge the EPA, the FDA, and Congress to make crucial changes in the way pesticides are regulated, in order to protect children against pesticides in food. If you'd like to write to other officials as well, page 67 of the Tools section includes more suggestions and addresses.

Dear Commissioner Young:

I am very concerned about dangerous pesticide residues in my family's food. On behalf of my family and other American families, I ask you to immediately make whatever changes are necessary for the FDA to do a much better job of detecting pesticides in our country's food supply, and to ensure that no food is sold that contains illegal pesticide residues.

Signed _____

Address _____

_____

Dear Administrator Reilly:

I am very concerned about dangerous pesticide residues in my family's food, particularly in the fruits and vegetables that I feed my children. A recent study by the Natural Resources Defense Council found that children are exposed to pesticides in food at greater levels than adults, at the time in their lives when they may be most vulnerable to pesticides' hazards. I ask that you take steps immediately to replace the existing, outdated limits on pesticide residues in food with new limits that *truly* protect the public–especially children.

Signed _____

Address _____

_____

Dear Chairman Waxman:

I am very concerned about the findings of a recent study by the Natural Resources Defense Council, showing that our children are being exposed to dangerous levels of pesticides in their food. I ask that Congress immediately reform the Federal Food, Drug and Cosmetic Act to establish strict, health-based legal limits that fully protect children from pesticides in food, and to stengthen monitoring for pesticide residues to enforce those standards effectively.

Signed _____

Address _____

_____

Frank E. Young, M.D.
Commissioner
Food and Drug Administration
Department of Health and Human Services
5600 Fishers Lane
Rockville, MD 20857

Mr. William Reilly
Administrator
Environmental Protection Agency
401 M Street, S.W.
Washington, D.C. 20460

The Honorable Henry Waxman, Chairman
House Energy and Commerce Committee,
Subcommittee on Health and the Environment
Room 2424, RHOB
Washington, D.C. 20515

Dear Representative Madigan:

I am very concerned about the findings of a recent study by the Natural Resources Defense Council, showing that our children are being exposed to dangerous levels of pesticides in their food. I ask that Congress immediately reform the Federal Food, Drug and Cosmetic Act to establish strict, health-based legal limits that fully protect children from pesticides in food, and to stengthen monitoring for pesticide residues to enforce those standards effectively.

Signed _____

Address _____

_____

Dear Chairman Kennedy:

I am very concerned about the findings of a recent study by the Natural Resources Defense Council, showing that our children are being exposed to dangerous levels of pesticides in their food. I ask that Congress immediately reform the Federal Food, Drug and Cosmetic Act to establish strict, health-based legal limits that fully protect children from pesticides in food, and to stengthen monitoring for pesticide residues to enforce those standards effectively.

Signed _____

Address _____

_____

Dear Senator Hatch:

I am very concerned about the findings of a recent study by the Natural Resources Defense Council, showing that our children are being exposed to dangerous levels of pesticides in their food. I ask that Congress immediately reform the Federal Food, Drug and Cosmetic Act to establish strict, health-based legal limits that fully protect children from pesticides in food, and to stengthen monitoring for pesticide residues to enforce those standards effectively.

Signed _____

Address _____

_____

The Honorable Edward Madigan,
Ranking Minority Member
House Energy and Commerce Committee,
Subcommittee on Health and the Environment
Room 2424, RHOB
Washington, D.C. 20515

The Honorable Edward M. Kennedy
Chairman, Senate Labor and
Human Resources Committee
SD-428
Washington, D.C. 20510

The Honorable Orrin G. Hatch,
Ranking Minority Member
Senate Labor and Human Resources Committee
SD-428
Washington, D.C. 20510

How to join Mothers and Others for Pesticide Limits

The Mothers and Others campaign was established to give a powerful, unified voice to concerns about pesticides in our children's food. Mothers and Others will be working to call widespread public attention to this problem, to press for important changes in the way pesticides are regulated and in the way the country's food supply is grown, and to ensure that safe, pesticide-free food is widely available in stores across the country.

We need to have many American families join us in our efforts. Together, we can have an impact on federal law and on the marketplace, so that our children are protected against pesticides in their food. We'll keep you informed of our successes and of new developments concerning the safety of children's food in our quarterly newsletter and in other alerts and updates.

Use this form to join us today. And please let us know that you have mailed in your postcards to Congress, the EPA, and the FDA.

Mothers and Others
FOR PESTICIDE LIMITS

Boynton

A PROJECT OF THE
NATURAL RESOURCES DEFENSE COUNCIL

___ **Yes,** I'd like to join NRDC's Mothers and Others for Pesticide Limits campaign. My tax deductible contribution of $15 or more entitles me to receive the Mothers and Others quarterly newsletter, as well as other timely information and alerts on the problem of pesticides in children's food.

Amount enclosed:
$15___ $25___ $100___ $_____ (other)

___ I've mailed in my postcards to Congress, the EPA, and the FDA.

___ I'd like additional copies of _For Our Kids' Sake_.

Amount enclosed:
($7.95 each includes
postage and handling) $_____

Total amount enclosed: $_____

Please make check payable to Mothers and Others.
Return to: Mothers and Others
P.O. Box 96641
Washington, D.C. 20090

Name
_____

Address
_____

City   State   Zip

A copy of NRDC's last financial report filed with the New York Department of State may be obtained by writing to: New York Department of State, Office of Charities Registration, Albany, NY 12231, or to NRDC.

MMY

Green Beans

Dimethoate *Cygon, Rogon*	Residues are systemic. However, washing, peeling, cooking, or heat processing reduced residues in various studies.
Methamidophos *Monitor*	Residues are systemic and probably cannot be removed with washing.
Endosulfan *Thiodan*	Residues remain primarily on the produce surface; however, endosulfan metabolites may be systemic. Peeling, cooking, or heat processing may reduce residues slightly. No information on removal with water.
Acephate *Orthene*	Residues are systemic and probably cannot be removed with washing. Cooking or canning may reduce residues.
Chlorothalonil *Bravo*	Residues remain primarily on the produce surface; however, chlorothalonil metabolites may be systemic. Washing or cooking reduces residues.

Lettuce

Mevinphos *Phosdrin*	Residues are systemic and probably cannot be removed with washing.
Endosulfan *Thiodan*	Residues remain primarily on the produce surface; however, endosulfan metabolites may be systemic. No information on removal with water.
Permethrin *Ambush, Pounce*	Residues remain primarily on the produce surface. Washing with detergent will reduce residues; plain water may not.
Dimethoate *Cygon, Rogon*	Residues are systemic. However, washing and peeling reduced residues in various studies.
Methomyl *Lannate*	Residues are systemic and probably cannot be removed with washing.

Onions

DCPA *Dacthal,* *Chlorthaldimethyl*	Some evidence that residues are systemic. For a preemergence herbicide, the relatively frequent findings of residues indicate a fair degree of persistence. No information on removal with water.
DDT*	Residues remain primarily on the produce surface, although residues may be absorbed into the peel. Washing and peeling reduce residues in root crops, but cooking won't reduce residues once they have been absorbed into the plant tissue.
Ethion *Ethanox, Ethiol,* *Rhodocide*	Residues remain primarily on the produce surface. Washing or processing may reduce residues.
Diazinon *Spectrocide,* *Sarolex*	Residues remain primarily on the produce surface. Cooking or heat processing may reduce residues. No information on removal with water.
Malathion *Cythion*	Residues remain primarily on the produce surface, but may be absorbed into the peel. Washing with detergent reduces residues more than plain water. Peeling, cooking, or heat processing will reduce residues.

*Pesticides in bold type are especially hazardous.

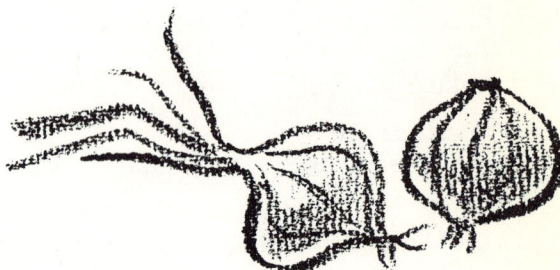

Oranges

Methidathion *Supracide, Somonil*	Residues remain primarily in the peel of citrus fruit. No information on removal with water.
Chlorpyrifos *Dursban*	Residues remain primarily on the produce surface. No information on removal with water.
Ethion *Ethanox, Ethiol, Rhodocide*	Residues remain primarily on the produce surface. Washing or processing may reduce residues.
Parathion *Phoskil*	Residues remain primarily on the produce surface. Washing, peeling, cooking, or heat processing may reduce residues slightly.
Carbaryl *Sevin*	Residues remain primarily on the produce surface. Washing, peeling, or cooking will reduce residues.

Peaches

Dicloran *DCNA, Botran*	Residues remain on surface following foliar treatment but are absorbed and translocated to edible tissue, following soil treatment. Incorporation of dicloran into wax formulations reduces the effectiveness of washing. Washing, peeling, cooking, or heat processing may reduce residues.
Captan *Merpan, Orthocide*	Residues remain primarily on the produce surface. However, the metabolite THPI, a suspected carcinogen, may be systemic. Washing, cooking, or heat processing will reduce residues.
Parathion *Phoskil*	Residues remain primarily on the produce surface. Washing, peeling, cooking, or heat processing may reduce residues slightly.
Carbaryl *Sevin*	Residues remain primarily on the produce surface. Washing, peeling, or cooking will reduce residues.
Endosulfan *Thiodan*	Residues remain primarily on the produce surface; however, endosulfan metabolites may be systemic. Peeling, cooking, or heat processing may reduce residues slightly. No information on removal with water.

Fact:
By the time FDA laboratories can analyze food samples for illegal pesticide residues, the contaminated food has most likely already been sold and eaten.

47

Pears

Azinphos-methyl *Guthion*	Residues remain primarily on the produce surface. Washing, cooking, or heat processing will reduce residues.
Cyhexatin *Plictran*	Residues remain primarily on the produce surface. Cyhexatin does not readily dissolve in water. Therefore, plain water washing may not reduce residues, but peeling may help. Residues in processed fruits are greater than in fresh fruits.
Phosmet *Imidan*	Residues remain primarily on the produce surface. Washing or cooking will reduce residues.
Endosulfan *Thiodan*	Residues remain primarily on the produce surface; however, endosulfan metabolites may be systemic. Peeling, cooking, or heat processing may reduce residues slightly. No information on removal with water.
Ethion *Ethanox, Ethiol, Rhodocide*	Residues remain primarily on the produce surface. Washing or processing may reduce residues.

Potatoes

DDT	Residues remain primarily on the produce surface, although residues may be absorbed into the peel. Washing and peeling reduce residues in potatoes, but cooking won't reduce residues once they have been absorbed into the plant tissue.
Chlorpropham *CIPC*	Residues are primarily systemic and probably cannot be removed by washing.
Dieldrin	Residues remain primarily on the produce surface, although there is some evidence that residues are systemic in root crops. No information on removal with water. Peeling or cooking may reduce residues.
Aldicarb *Temik*	Residues are systemic and probably cannot be removed with washing. Cooking or heat processing may reduce residues.
Chlordane *Octachlor, Velsicol 1068*	Residues remain primarily on the produce surface, although there is some evidence that residues are systemic in root crops. No information on removal with water.

Tip:
Waxes can't be washed off produce, and they can seal in pesticides. Some waxes may be obvious, such as on cucumbers. But many other fruits and vegetables are frequently waxed as well, including avocados, bell peppers, cantaloupes, eggplants, grapefruits, lemons, limes, melons, oranges, parsnips, passion fruits, peaches, pineapples, rutabagas, squashes, sweet potatoes, tomatoes, and turnips.

Spinach

Endosulfan
Thiodan

Residues remain primarily on the produce surface; however, endosulfan metabolites may be systemic. Cooking or heat processing may reduce residues slightly. No information on removal with water.

DDT

Residues remain primarily on the produce surface, although residues may be absorbed into the peel. Washing, cooking, and commercial processing will reduce residues to some extent in spinach.

Methomyl
Lannate

Residues are systemic and probably cannot be removed with washing. Residues are stable at freezing temperatures; however, cooking or heat processing will reduce residues.

Methamidophos
Monitor

Residues are systemic and probably cannot be removed with washing. Residues remain primarily in the outer leaves, so stripping these tough layers in spinach may reduce residues.

Dimethoate
Cygon, Rogon

Residues are systemic. However, washing, cooking, and heat processing reduced residues in various studies.

Strawberries

Captan
*Merpan,
Orthocide*

Residues remain primarily on the produce surface. However, the metabolite THPI, a suspected carcinogen, may be systemic. Washing, cooking, or heat processing will reduce residues.

Vinclozolin
Ronilan

Residues remain primarily on the produce surface, but some evidence that residues can be absorbed. Cooking or heat processing will reduce residues.

Endosulfan
Thiodan

Residues remain primarily on the produce surface; however, endosulfan metabolites may be systemic. Cooking or heat processing may reduce residues slightly. No information on removal with water.

Methamidophos
Monitor

Residues are systemic and probably cannot be removed with washing.

Methyl Parathion
Folidol M, Metacide

Residues remain primarily on the produce surface, but some evidence that residues can be absorbed. No information on removal with water.

49

Sweet Potatoes

Dicloran *DCNA, Botran*	Residues remain on surface following foliar treatment, but are absorbed and translocated to edible tissue following soil treatment. Incorporation of dicloran into wax formulations reduces the effectiveness of washing. Washing, peeling, cooking, or heat processing may reduce residues.
Phosmet* *Imidan*	Residues remain primarily on the produce surface. Washing or cooking will reduce residues.
DDT	Residues remain primarily on the produce surface, although residues may be absorbed into the peel. Washing and peeling reduce residues in root crops, but cooking won't reduce residues once they have been absorbed into the plant tissue.
Dieldrin	Residues remain primarily on the produce surface, although there is some evidence that residues are systemic in root crops. No information on removal with water. Peeling or cooking may reduce residues.
BHC *HCH, 666,* *Hexachlor*	Residues remain primarily on the produce surface, although there is some evidence that residues are systemic in root crops. No information on removal with water.

*Pesticides in bold type are especially hazardous.

Tomatoes

Methamidophos
Monitor

Residues are systemic and probably cannot be removed with washing.

Chlorpyrifos
Dursban

Residues remain primarily on the produce surface. No information on removal with water. Dried fruits showed higher levels of residues than on fresh fruits because of the concentration effect of dehydration.

Chlorothalonil
Bravo

Residues remain primarily on the produce surface; however, chlorothalonil metabolites may be systemic. Washing or cooking reduces residues.

Permethrin
*Ambush,
Pounce*

Residues remain primarily on the produce surface. Washing with detergent will reduce residues; plain water may not. Processing tomatoes causes residues to concentrate at far greater levels than in fresh produce.

Dimethoate
*Cygon,
Rogon*

Residues are systemic. However, washing, peeling, cooking, or heat processing reduced residues in various studies.

Watermelon

Methamidophos
Monitor

Residues are systemic and probably cannot be removed with washing.

Chlorothalonil
Bravo

Residues remain primarily on the produce surface; however, chlorothalonil metabolites may be systemic. Washing reduces residues.

Dimethoate
*Cygon,
Rogon*

Residues are systemic. However, washing and peeling reduced residues in various studies.

Carbaryl
Sevin

Residues remain primarily on the produce surface. Washing and peeling will reduce residues.

Captan
*Merpan,
Orthocide*

Residues remain primarily on the produce surface. However, the metabolite THPI, a suspected carcinogen, may be systemic. Washing, cooking, or heat processing will reduce residues.

Tip:
Lettuce, tomatoes, and zucchini are relatively easy to grow. When starting a home garden, be sure the soil you're using is not contaminated by previous pesticide use.

Tips on organizing locally for pesticide-free food

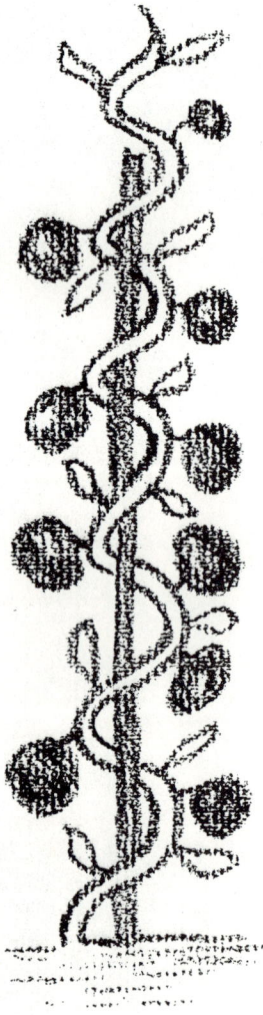

The real key to protecting our children against pesticides in food will be for large numbers of people to become involved in this issue. Whatever you can do alone about the pesticides problem, you can do even more effectively if others join you. There's plenty for a group of people to do—for instance, convincing supermarkets to stock safe food, working with local farmers to reduce pesticide use, and organizing local lobbying efforts to urge congressional representatives to enact important reforms. Here are some tips for getting started:

- There's nothing mysterious about doing this kind of organizing. **It almost always starts, quite simply, with talking.** You mention the problem of pesticides in children's food to a neighbor or a friend; he or she talks to another friend about it, and soon you have the core of a local group.

- **Your next step is to identify your "allies."** Think of *existing* groups that are likely to be interested in this issue, such as environmental and consumer organizations, health food stores and food co-ops, farmers' groups, civic clubs, and PTAs and other organizations concerned with children's issues. Contact these groups to see if you can attend some of their meetings to talk about the problem of pesticides in children's food and to invite them to join with you in doing something about it.

- **Hold an organizing meeting.** Have hand-out material about the issue available, and pass around a sign-up sheet for people's names, addresses, phone numbers, and affiliations. Some projects that you might want to discuss or undertake include: lining up volunteers to attend local organizations' meetings to talk about pesticides in children's food; starting letter-writing campaigns to Congress, the EPA, FDA, and USDA, as well as writing letters to the editor of your local newspaper; compiling information on local sources of organically-grown produce; and planning how to approach local supermarkets about stocking pesticide-free food.

How to get organic food into your supermarket

Your shopping dollars count!

Supermarket managers care about what consumers want. If you go elsewhere for pesticide-free produce, the food buyers at your local supermarket will take note.

Right now, without the incentive of customer demand, most stores don't offer shoppers the choice of organically-grown food. But you and other consumers, speaking with informed concern, can convince a supermarket buyer to offer the choice of fresh produce grown without pesticides.

3 basic steps to get organic food into your supermarket

1. Go to the produce section of your supermarket and ask to see the produce manager. Tell him or her that you are looking for *certified* organically-grown produce (for more information on certification of organic food, see page 59). If the store doesn't offer it, say that you are concerned about pesticide residues in your produce and that you would like the choice of buying organically-grown food. If you make the point that you will purchase your produce elsewhere unless you have a choice, the produce manager is likely to listen.

2. Write to the corporate management of your supermarket to express your concern about pesticide residues in food, and your desire for certified organic produce. Mention that many supermarkets are beginning to stock organically-grown food, and that you will shop where it is available. You can also circulate a petition to be signed by supermarket customers, requesting that the store offer pesticide-free food. There's a sample petition on page 55.

You can get the address for your supermarket's corporate office from the local supermarket manager. Be sure to leave a copy of your letter with the local manager, and encourage your friends to write as well. Most supermarkets rarely get mail from their customers, so just a few letters can make a difference.

3. Help your supermarket management establish direct contact with growers and wholesalers of organic foods. Supermarket produce buyers might not know where to obtain certified organic food. For listings of distributors servicing your area, see the resources for supermarkets below. In addition, groups that certify

Adapted with permission from Organic Farms, Inc., 10714 Hanna St., Beltsville, MD 20705, (301) 595-5151.

organically-grown food, such as the California Certified Organic Farmers (CCOF), can also identify suppliers and wholesalers of organic food. CCOF has two helpful references: a list of organic growers in California who will sell directly to supermarkets, and a list of wholesale and retail sources of organic produce. The lists are available for $5.00 from CCOF, P.O. Box 8136, Santa Cruz, CA 95061.

The myth of "not enough supply"

Although organic food and produce isn't as readily available as foods grown commercially with pesticides, the demand for organic food has not yet exceeded the available supply. Successful organic farming techniques have been available for many years. The expanded use of alternative pest control methods will depend, in part, on a decision by consumers to demand an end to the current dependence on chemical pesticides. Once growers perceive a larger demand for organic food, they will recognize the economic feasibility of switching to safer pest control techniques. Farmers will be much more willing to change their agricultural methods if they know that consumers will support their efforts. So your support of organic farming will mean that increased amounts of organic food will be available—which in turn will bring about lower prices.

Some resources for supermarkets and consumers:

Organic Wholesalers Directory and Yearbook, California Action Network, 1988. (Available from CAN, P.O. Box 464, Davis, CA 95617, $19.00, plus $2.00 for shipping and handling.) An annual guide for farmers who want to sell their organic commodities through a wholesaler, and for retailers looking for wholesale markets of organic food.

The Organic Network, Jean Winter, 1984. (Available from Eden Acres, Inc., 12100 Lima Center Road, Clinton, MI 49236, $15.00) A directory listing organic growers and buyers, as well as individuals, farmers, and businesses selling organically-grown food.

Tip:
It may be better to buy produce grown in the drier climates of the West and Southwest, since humidity forces higher use of fungicides—90 percent of which (by total tonnage) are carcinogenic.

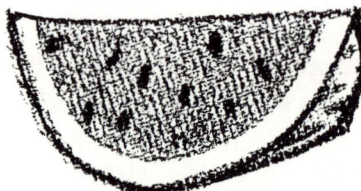

Petition for pesticide-free food

Here's a sample petition that you can give to your supermarket management requesting that the store offer pesticide-free produce for sale. You can copy this page, or you can write your own petition using the information contained in this book. In addition to delivering signed petitions to your local supermarket, be sure to send copies to the store's corporate headquarters.

Dear supermarket manager:

As regular customers in your store, we are concerned about pesticide residues in food. In particular, we are disturbed by the findings of a recent study by the Natural Resources Defense Council showing that our children are being exposed to dangerous levels of pesticides in their food. Therefore, we request that your supermarket locate suppliers of pesticide-free food and make it available in your store. Other supermarkets across the country are offering certified organic produce for sale. We urge you to do the same.

Signed   **Address**

_____

_____

_____

_____

_____

_____

_____

_____

Working with farmers for safer produce

• Ed Sills grows organic yellow corn, popcorn, and rice on a 1,400-acre farm in Pleasant Grove, California. By rotating his regular crops with nitrogen-rich purple vetch, Sills *increases* the fertility of his soil while *decreasing* insect infestations. Although his organic crop yields are 25 percent lower than the yields of his few remaining chemically-grown crops, the organic crops earn twice as much because his input expenses are down considerably.

• Steve Pavich's 1,270-acre farm in Delano, California is the single largest organically-grown table grape business in the world. Pavich uses biological insect controls, and enriches the soil with cover crops and manure.

• For more than 17 years, the Schroeder family has been farming 222 acres of corn, soybeans, grains, and livestock in Columbus Grove, Ohio without the use of pesticides and fertilizers. The Schroeders use mechanical cultivation to control weeds, and alternate crops that deplete soil nutrients with soil-building crops. Their corn yields are slightly lower than their neighbors', but this is offset by lower costs.

• North Dakota farmers Fred and Janet Kirshenmann grow sunflower, millet, and other cereal grains, and flax and alfalfa for feeding cattle. Their success in organic farming has allowed them to expand their farm from 480 to 2,100 acres in 12 years.

• Dick and Sharon Thompson quit using chemicals on their 300-acre hog and beef farm in central Iowa back in 1967. Instead, they now use various methods of mechanical weed control and cover crops to give their crops every possible advantage over weeds. And they do it at savings of $23 per acre for soybeans and $18.50 an acre for corn.

"Success stories" like these abound these days, as more and more farmers are turning—profitably—to low-chemical or no-chemical farming methods. In California alone, for instance, the total acreage under organic production more than doubled in the past two years, from 10,000 acres to nearly 26,000. Nationwide, the market for food grown without chemicals has increased 20-fold since the early 1970s, to as much as $10 billion a year.

There are certainly compelling reasons for farmers to make the switch, including concerns about the health and economic costs of increased pesticide use. In fact, since the 1970s there has been a growing trend toward a "sustainable" form of agriculture that preserves resources while protecting the health of consumers and the environment, *and* maintaining a steady income for farmers. This "Low-Input Sustainable Agriculture," or "LISA," involves farming methods that range from targeted, more efficient pesticide use to totally organic systems that eliminate the use of pesticides and fertilizers altogether. Specifically, low-input practices might include *integrated pest management*, which relies on careful monitoring to detect when pesticide applications will have the maximum effect; *crop rotation*, which helps break the reproductive cycle of most insects; *biological controls*, which assault pests with natural predators, parasites, and pathogens; and the use of *cover crops* and plant varieties that resist pests.

If the trend toward safer farming is to continue and to accelerate, though, farmers are going to need support from consumers and the government. There are several things that you can do. First, you can buy directly from local farmers who grow crops organically, or are in transition from high-chemical to no-chemical farming. You can also help those farmers to expand their markets for organic produce, by putting supermarkets and other retail outlets in touch with them. You can write to Congress and the USDA, to urge the government to provide farmers with incentives to switch to pesticide-free farming. And you can give farmers useful information about the advantages of safer farming methods, as well as sources of technical help and ways to network with other organic farmers.

Some resources for farmers:

Institute for Alternative Agriculture
9200 Edmonston Rd., Suite 117
Greenbelt, MD 20770
(301) 441-8777

A clearinghouse on alternative agriculture, and a "voice" supporting research and education. Publishes a monthly newsletter; organizes an annual policy symposium.

Regenerative Agriculture Association
222 Main St.
Emmaus, PA 18049
(215) 967-5171

Publishes *The New Farm* magazine, books, and other special publications; sponsors field days; operates readers' service that finds answers to farmers' specific questions or makes referrals to

Tip:
The government recommends washing fruits and vegetables for three minutes, and rinsing well. Heating or cooking produce may help reduce some pesticide residues.

researchers or other farmers via *Farmers Own Network for Extension*; is dedicated to putting people, profit, and biological permanence back into agriculture. Affiliated with the Rodale Research Center, R.D. 1, Box 323, Kutztown, PA 19530.

Organic Foods Production Association of North America (OFPANA)
P.O. Box 31
Belchertown, MA 01007
(413) 323-6821

A trade association that establishes organic guidelines and promotes organic certification. Publishes "Guidelines for the Organic Foods Industry (cost $10, plus $2.50 handling) and "Laboratory Testing and the Production and Marketing of Certified Organic Foods" (cost $4). Also produces a variety of promotional materials.

Organic Crop Improvement Assocation (OCIA)
3185 Twp. Rd. 179
Bellefontaine, OH 43311
(513) 592-4983

An international certifying group with 27 chapters throughout North America and several foreign countries.

Farm Verified Organic (FVO)
P.O. Box 45
Redding, CT 06875
(203) 544-9896

An international non-profit organization founded in 1984 to certify organic food. Offers educational, marketing, and technical services to more than 200 members farming some 250,000 acres worldwide. Also licenses its trademark to food companies in 15 countries.

Sources of organic produce

In addition to the mail-order sources of organic food listed on pages 63–66, Americans For Safe Food publishes a state-by-state directory listing about 100 organic food mail-order retail suppliers. To receive a copy of this directory, send 50 cents for handling, and a self-addressed, stamped business envelope (50 cents postage) to Americans For Safe Food, 1501 16th St., N.W., Washington D.C. 20036.

A note about "certification"

Organic farmers have recognized the need for a respected verification process that can offer consumers assurance that food labeled "organic" meets certain standards. They have responded by forming independent certifying agencies. These agencies add to the credibility of the true organic farmer and increase the integrity of the marketplace in general.

Most of these certifying organizations offer technical assistance to farmers as well. Some hire third-party inspectors to prevent any mischief in the certification process. When a product is labeled "certified organic," it indicates that the farm where the food was grown has met certain standards of organic farming. Generally, an inspector visits each farm that applies for certification to examine the growing process.

"Organic" farming is generally defined as farming that minimizes or avoids the use of synthetic fertilizers, pesticides (including herbicides, insecticides, and fungicides), **growth regulators**, and feed additives. This is a limited definition, as many certifying agencies have more extensive criteria that a farmer must meet.

There are five primary methods used to verify organic products: questionnaires, affidavits, inspection reports, laboratory analyses, and audits. The following issues can be used to evaluate a particular certification process:

• **Testing and inspection.** Most certifiers subject applicants to a soil test or analysis. Some do residue testing as well. You should ask about ongoing inspections, i.e., how often does the certifying agency do spot checks on farms certified as organic? Are all inspections made by a third party to assure objectivity?

• **Application process.** Most certifiers require farmers to submit an application in order to be considered for certification. Some require a sworn and notarized affidavit for the first year or longer. In

Reprinted with permission from Americans For Safe Food, 1501 16th St., N.W., Washington D.C. 20036 all cases, the farmer must supply the certifier with accurate information about his past and present farming practices.

• **Philosophy.** Many certifiers believe strongly in the philosophy of sustainability. Sustainable agriculture generally means a system of ecological soil management practices and low-input farming leading to a self-sustained ecosystem that does not require artificial or chemical treatment. Sustainable farming is not necessarily organic farming, but it is a vast improvement over conventional chemical farming.

• **Transitional farmer.** Most certification programs require some period of transition before a farm that has been under conventional management can be certified as organic. Certifiers usually require delays of at least one year from the last use of prohibited fertilizers and two to three years from the last use of prohibited pesticides.

Some programs do certify transitional farmers as such, partially to encourage more conventional farmers to make the switch to organic techniques. Other certifiers, however, do not see transitional certifications as benefitting the consumer.

• **Working toward a national standard.** A common definition for "organic" is badly needed to lend more consistency and uniformity to the term. The Organic Foods Production Association of North America (OFPANA) has adopted national guidelines that amount to a first step in that direction. OFPANA is a trade association consisting of organic producers, retailers, and certifying agencies. It considers itself the certifier of certifiers. It will offer its endorsement to certifying agencies whose standards meet the OFPANA criteria.

Because there is no national standard as yet for the certification of organic food, the methods and standards established by certifying agencies do differ. Many of these agencies have published pamphlets or other documents that describe their certification standards and requirements, and they will be glad to send them to you. Although most of these organizations deal most often with farmers, some have consumer memberships as well. In any event, organic farmers are your natural allies. Their organizations should be valuable sources of information, energy, and support for your efforts. In return, your efforts should include a search for helpful information on organic farming or on any grants or support programs available to farmers considering the switch to organic farming.

The following are some of more than 40 certifying agencies in the United States. The list also includes several organizations that can help you locate wholesalers and retailers of organic products.

Fact:
Some 60 to 80 percent of the pesticides applied to citrus, and 40 to 60 percent of pesticides applied to tomatoes grown for processing, are used for cosmetic purposes.

California Action Network
P.O. Box 464
Davis, CA 95617
(916) 756-8518

California Certified Organic Farmers (CCOF)
P.O. Box 8136
Santa Cruz, CA 95061
(408) 423-2263

Demeter Association
West of the Mississippi:
4214 National Avenue
Burbank, CA 91505
(818) 363-7312

Demeter Association
East of the Mississippi:
P.O. Box 6606
Ithaca, NY 12851

Farm Verified Organic Program (FVO)
Mercantile Development Inc.
274 Riverside Ave.
P.O. Box 2747
Westport, CT 06880
(203) 226-7803

Maine Organic Farmers' & Gardeners' Association
Jay Adams
P.O. Box 2176
Augusta, ME 04330
(207) 622-3118

Natural Organic Farmers Association (NOFA)
P.O. Box 335
Antrim, NH 03440
(603) 588-6668

NOFA – Vermont
15 Barre Street
Montpelier, VT 05602
(802) 223-7222

NOFA – Massachusetts
21 Great Plain Ave.
Wellesley, MA 02181

NOFA – Connecticut
100 Rose Hill Road
Branford, CT 06405

NOFA – New York
P.O. Box 454
Ithaca, NY 14851

Ohio Ecological Food and Farm Association
7300 Bagley Rd.
Mt. Perry, OH 43769
(614) 448-3951

Organic Crop Improvement Association (OCIA)
125 West Seventh St.
Wind Gap, PA 18091
(215) 863-6700

Organic Food Network
c/o American Fruitarian Society
6600 Burleson Road
P.O. Box 17128
Austin, TX 78760-7128
(512) 385-2841

Organic Foods Production Association of North America (OFPANA)
c/o Judith Gillan
P.O. Box 31
Belchertown, MA 01007
(413) 323-6821

Organic Growers and Buyers Association
P.O. Box 9747
Minneapolis, MN 55440
(612) 674-8527

Organic Growers of Michigan
c/o Lewis King
3031 White Creek Rd.
Kingston, MI 48741
(517) 683-2573

Texas Department of Agriculture Organic Certification Program
P.O. Box 12847
Austin, TX 78711
(512) 463-9883

Tilth Producers' Cooperative
c/o Yvonne Frost
P.O. Box 218
Tualatin, OR 97062

Tilth Producers' Cooperative
1219 East Sauk Road
Concrete, WA 98237
(206) 853-8449

Mail order sources of organic food

Organic Foods

Organic Foods Express
11003 Emack Rd.
Beltsville, MD 20705
(301) 937-8608

full line of fruits and vegetables – flour, grains, seeds, beans, nuts

Rising Sun Organic Produce
Box 627 I-80
Milesburg, PA 16853
(814) 355-9850

wide variety of fruits and vegetables, dried fruit, nuts, beans, grains and meats

Krystal Wharf Farms
R.D. 2, Box 191-A
Mansfield, PA 16933
(717) 549-8194

grains, seeds, beans, nuts, dried fruit and pasta

Walnut Acres
Penns Creek, PA 17862
(717) 837-0601

full line of cereals, flours, grains, baked goods, soups, oils, peanut butter, canned and dried vegetables

The Green Earth
2545 Prairie Ave.
Evanston, IL 60201
(800) 322-3662

wide variety of fruits and vegetables

Stanley and Marina Jacobson
1505 Doherty
Mission, Texas 78572
(512) 585-1712

grapefruit and oranges

Earth's Best Baby Food
P.O. Box 887
Middlebury, VT 05753
(800) 442-4221

Not a retail mail order company, but products are available through Walnut Acres, above; or call Earth's Best for the distributor nearest you.

Millstream Natural Health Supplies
1310-A E. Tallmadge Ave.
Akron, OH 44310
fruits, vegetables, cheese and nuts

Starr Organic Produce, Inc.
P.O. Box 561502
Miami, FL 33256
(305) 262-1242
oranges and grapefruit

Blue Heron Farm
P.O. Box 68
Runsey, CA 95679
(916) 796-3799
oranges, almonds and walnuts

Ecology Sound Farms
42126 Rd. 168
Orosi, CA 93647
fruit

Hill and Dale Farms
West Hill – Daniel Davis Rd.
Putney, VT 05346
(802) 387-5817
apples

Green Knoll Farm
P.O. Box 434
Gridley, CA 95948
kiwi

Weiss's Kiwifruit
594 Paseo Companeros
Chico, CA 95926
kiwi

Deer Valley Farm
R.D. 1
Guilford, NY 13780
pasta, flour, grains, seeds and fruit

Hawthorne Valley Farm
R.D. 2, Box 225A
Ghent, NY 12075
(518) 672-7500
cheese, bread, granola

Berkshire Mountain Bakery
P.O. Box 785
Housatonic, MA 01236

bread

Baldwin Hill Corporation
Baldwin Hill Rd.
Phillipston, MA 01331

bread

Bread Alone
Rt. 28
Boiceville, NY 12412

bread

Community Mill and Bean
R.D. #1, Rt. 89
Savannah, NY 13146
(315) 365-2664

flours, grains and beans

Mountain Ark Trading Company
120 South East Ave.
Fayetteville, AK 72701
(800) 643-8909

grains, beans, flours, dried fruits, nuts and oils

Gold Mine Natural Food Co.
1947 30th St.
San Diego, CA 92102
(619) 234-9711

grains, beans, macrobiotic foods

Fiddler's Green Farm
R.F.D. 1, Box 656
Belfast, ME 04915

coffee and whole grain baking mixes

Terra Nova
P.O. Box 1647
Ojai, CA 93023

coffee

Natural Beef Farms Food Distribution Co.
4399-A Henninger Ct.
Chantilly, VA 22021
(703) 631-0881

beef, chicken, turkey, lamb, veal and seafood

Four Chimneys Farm Winery
R.D., #1, Hall Rd.
Himrod-on-Seneca, NY 14842
wine

George P. Nassopoulos
195 Whiting St.
Hingham, MA 02043
(617) 749-1866
olive oil

Garden Supplies

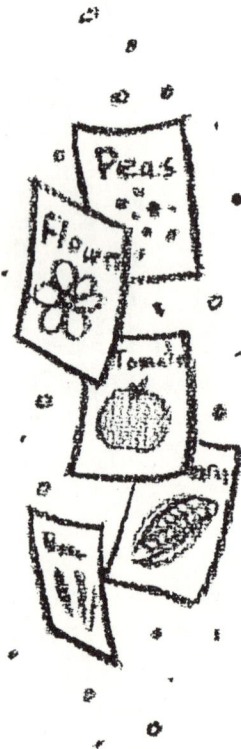

Butterbrooke Farm
78 Barry Rd.
Oxford, CT 06483
untreated vegetable seeds

Pinetree Garden Seeds
New Gloucester, ME 04260
(207) 926-3400
untreated vegetable seeds

Gardener's Supply
128 Intervale Rd.
Burlington, VT 05401
natural and organic pest controls

Natural Gardening Research Center
Hwy. 48, P.O. Box 149
Sunman, Indiana 47041
catalog of natural and organic gardening supplies

Richters
Goodwood
Ontario, Canada LOC 10A
French and Mexican natural nematode control marigold seeds

Organic Seed Savers Exchange
c/o Kent Whealy
Decorah, Iowa
organic garden seed bank and exchange

Tips on letter-writing

The best way to write to government and elected officials is to use your own words to express your personal concerns and say what action you want the official to take. Letters should be type-written or neatly handwritten. In addition to writing letters to the following officials and elected representatives, if you have time, you may want to write to your supermarket management to ask them to stock organically-grown produce (see page 53); the editor of your local newspaper (the letters page is one of the most popular parts of the newspaper, so it's a good way to call attention to an issue); state agricultural officials and health officials to see if your state has any programs to deal with pesticides in food; and state legislators, to ask them to put pressure on Congress and federal officials for important reforms.

Here are suggestions about what officials to write to, and what to emphasize in your letters (you can also refer to the recommendations on pages 28 in the **Solutions** section for ideas on what to write):

- **Your own U.S. Senator and Representatives.** Ask them to work to establish stricter controls on pesticides in food; increased research on alternatives to high-chemical farming; and incentives to farmers to switch to safer farming methods.

> The Honorable _____
> U.S. Senate
> Washington, D.C. 20510

> The Honorable _____
> U.S. House of Representatives
> Washington, D.C. 20515

- **The Chairmen and Ranking Minority Members of key Senate and House Committees.** They are:

> The Honorable Henry Waxman, Chairman
> House Energy and Commerce Committee,
> Subcommittee on Health and the Environment
> Room 2424, RHOB
> Washington, D.C. 20515

Tip:
You may find organically-grown produce for sale at your local farmers' market. Ask whether the produce is *certified* organically-grown.

The Honorable Edward Madigan, Ranking
Minority Member
House Energy and Commerce Committee,
Subcommittee on Health and the Environment
Room 2424, RHOB
Washington, D.C. 20515

The Honorable Edward M. Kennedy
Chairman, Senate Labor and
Human Resources Committee
SD-428
Washington, D.C. 20510

The Honorable Orrin G. Hatch,
Ranking Minority Member
Senate Labor and Human Resources Committee
SD-428
Washington, D.C. 20510

• **The EPA and FDA.** Tell them your concerns about pesticides in children's food and the inadequacies in their programs to protect the food supply from dangerous pesticides. Write to:

Frank E. Young, M.D.
Commissioner
Food and Drug Administration
Department of Health and Human Services
5600 Fishers Lane
Rockville, MD 20857

William Reilly
Administrator
Environmental Protection Agency
401 M Street, S.W.
Washington, D.C. 20460

• **The U.S. Department of Agriculture.** Write to ask that the U.S.D.A. become a major advocate for farming methods that will reduce the reliance on pesticides in growing food crops. Write to:

Clayton K. Yeutter
Secretary of Agriculture
U.S. Department of Agriculture
14th and Independence Avenue, S.W.
Washington, D.C. 20250

Recommendations for reform

Chapter Five: Recommendations for Congressional Action Necessary to Protect Preschoolers from Pesticides in Food

Residues of pesticides with known toxic effects are widely present in the food supply. In particular, these residues pose a serious health threat to preschoolers. This public health problem is largely the result of inadequate regulation on the part of the federal government. Immediate action is necessary to close the loopholes in EPA's and FDA's regulation of pesticide residues in food. Further, Congress must act to assist growers in reducing their use of pesticides. This chapter provides specific recommendations for congressional action in order to ensure the safety of the food supply.

Congress Must Require EPA to Protect the Public From Pesticide Residues In Food.
The Federal Food, Drug and Cosmetic Act (FFDCA) governs the types of pesticides, and the concentrations of residues, which can legally occur in our foods. Tolerances, or legal limits on pesticide residues, are set by EPA under the FFDCA. EPA will be collecting important new data under the 1988 amendments to the FIFRA program, but it is not clear when or how EPA will use these data to reestablish tolerances at new levels which will truly protect public health.

A minimum of ten fundamental EPA reforms is needed to ensure the safety of our food supply from pesticide residues.

EPA Should Rapidly Establish New, Protective Tolerances for All 23 Pesticides Which Are the Subject of NRDC's Study.
For all of 23 chemicals which are the subject of NRDC's study, EPA should rapidly set legal limits which ensure that any residues of these chemicals persisting in our food supply do not present a hazard to preschoolers. It should take no longer than one year for EPA to set new, protective tolerances for each of the 23 chemicals in this report.

EPA Should Swiftly Reassess and Reestablish Tolerances for All Pesticides for Which EPA has Requested or Received New Health and Safety Data. For Pesticides That are Only Now Being Studied for Health Effects, EPA Should Never Take Longer Than One Year After Receiving New Studies to Establish Appropriate New Tolerances.
Pesticides in food are legal up to the tolerance level. Yet, residues are not necessarily safe in those amounts, or, as this report illus-

Reprinted with permission from "Intolerable Risk: Pesticides in Our Children's Food," by B. Sewell, R. Whyatt, J. Hathaway, and L. Mott, NRDC, 1989

trates, even at levels well below tolerance. This situation has resulted, in part, because most pesticide tolerances were set in the absence of reliable and complete data on the effects of long-term exposure. In fact, many existing tolerances were established by USDA prior to EPA's existence and were set with great deference to the economic interests of growers. Because the health effects of chronic low-level exposure were generally unknown, tolerance levels often only reflected residue levels resulting from growers' historical uses. Tolerances were routinely set high, so that crops subjected to the most frequent and most concentrated permissible pesticide treatments could be sold legally. EPA inherited these old tolerances from USDA when it was created in 1970. Unfortunately, many of these old, high tolerances still remain on the books. Tolerances have not been systematically reassessed and revised as EPA receives new data indicating the potential of pesticides to cause cancer and other chronic effects.

This situation is exacerbated by the fact that even today, only a handful of the approximately 300 pesticides legally used on food have adequate, full data on their threat to human health from chronic dietary exposure. Amendments to the federal pesticide law enacted in September 1988 will require pesticide manufacturers to test their products fully in the next few years and EPA to reassess the safety of chemicals completely within nine years. Tolerance revision should occur immediately when EPA learns of new or higher risks than were assumed when the legal limit was initially set.

EPA Should Ensure That the Most Exposed and Most Vulnerable Members of Society—Infants and Children—are Adequately Protected From Pesticide Residues.

EPA's pesticide risk assessments and the resultant regulatory decisions assume an "average" diet. However, virtually every individual's diet deviates considerably from that of an average American's. Most people eat high quantities of certain foods and avoid other foods altogether. The consequence is that EPA's pesticide tolerances are designed to protect the theoretically average consumer—not the actual person who eats an apple, or two, a day. Consumers who eat higher quantities of any food than EPA considers to be average are not necessarily protected by EPA tolerances from developing cancers or other diseases caused by pesticides remaining on food.

As this study well illustrates, EPA's reliance on the average diet is particularly inappropriate for preschoolers. For example, the average preschooler consumes almost six times as much fruit, five times as much milk, and almost three and a half times as many grain products per unit of body-weight as the average adult woman. The difference for specific foods is often much greater. Clearly, children are not "average" consumers. Pesticide limits should be set to protect children and other "non-average" eaters. EPA should revise existing tolerances and establish new tolerances, based on the unique dietary intake of preschoolers and any other especially vulnerable or highly exposed subgroup of the population.

EPA Must Revise Exposure Estimates. Potential Exposure at the Legal Maximum—the Tolerance Level—Should be Assumed When EPA Conducts Risk Assessments.

EPA has not routinely lowered tolerances upon finding that the legally permissible concentrations of pesticide residues would cause a high risk of cancer or other adverse effects. EPA often dismisses these findings by stating that residues are normally below tolerance levels, but EPA often fails to lower the tolerances to what it has declared to be the "normal" level. Consequently, people may unwittingly consume foods that, although legal, contain pesticide levels which EPA's data demonstrate to be unsafe.

Legal food should be safe food. Without prohibitively expensive chemical analyses of all food, consumers cannot identify which foods bear high residues and which have lower, safer concentrations of pesticides. Therefore, EPA should ensure that consumption of food with residues at the legal maximum is safe. EPA should perform risk assessments based on the assumption that pesticide residues will be present at the maximum legal concentration. If growers or the food industry believe that such a risk assessment overstates the risk, they should demonstrate that actual residues are less and petition for a lower legal limit.

EPA Should Use the Most Current Information on Dietary Consumption Estimates to Set All New Tolerances and Should Revise All Existing Tolerances With These More Accurate Data.

EPA defends many old tolerances and even sets some new tolerances by reference to consumption estimates which tend to underestimate human dietary exposure. EPA's outdated consumption estimates ("Food Factors") presume that people eat no more than 7.5 ounces annually of many common foods including almonds, avocados, blueberries, cantaloupe, eggplant, honeydew melon, mushrooms, nectarines, and tangerines. These estimates greatly understate preschooler dietary consumption. For instance, preschooler intake of cranberries is 14 times greater than the comparable Food Factor, grape consumption is six times greater, apples and oranges approximately five times greater, apricots almost four times greater, and strawberries almost three times greater.

EPA now has more accurate data on food consumption based on a 1977 USDA dietary survey. EPA began utilizing the Tolerance

Assessment System (TAS), containing the 1977 data, in the fall of 1986 for extremely limited purposes. The TAS data have not been used to revise one single older tolerance initially established with the outdated Food Factors. Furthermore, the TAS data are now almost a decade old and American dietary habits have changed in the interim. For example, preschoolers' consumption of fruits and vegetables may have increased by 30%, and fruits alone by as much as 45%, between the 1977 and 1985 USDA dietary surveys. In addition, TAS can be used to examine the dietary intake of subgroups of the population, such as preschoolers. However, EPA primarily utilizes the TAS data on average adult consumption. EPA should immediately update TAS with the most current dietary intake data. All EPA tolerances, existing or new, should be established on the basis of TAS, with specific consideration of preschooler consumption patterns.

EPA Must Ensure That Cumulative Exposure to a Variety of Pesticides in Foods, in Drinking Water, and in Our Homes are Considered When Tolerance Levels are Set. It is the Cumulative Risk That the Person Actually Experiences, and That Risk Must be Minimized.

EPA's pesticide tolerances do not now ensure that exposures to all legally present pesticides in our food are safe. EPA's tolerance-setting is premised on the obviously erroneous assumption that we are exposed only to a single pesticide over our lifetimes. In fact, nearly 300 active ingredients in pesticides are registered for use on food and several may be present on a single strawberry or carrot. In 1984, NRDC analyzed pesticide residues in fresh produce purchased in grocery stores in San Francisco. NRDC found that two or more pesticide residues were present in 42% of the samples bearing identifiable residues. Further, in 1985 and 1986, the FDA found almost 7,000 residues of only 23 pesticides on approximately 4,000 samples of common fruits and vegetables.

But when EPA sets a tolerance level, EPA only calculates average risk from exposure to a single pesticide. The reality is that there may be several pesticides, which increase our risk of cancer or neurotoxic effects, present in a single meal. EPA currently fails to ensure that the cumulative risk of eating foods containing many carcinogenic or neurotoxic pesticides is safe.

EPA should also consider that consumers are exposed to many of the same pesticides used in foods in their drinking water and in their homes. An insecticide used to combat roaches in one's kitchen may be inhaled in small concentrations for months to come. When EPA permits the insecticide to be used on food crops, EPA should ensure that the total exposures from uses in the home, in the garden, and in agriculture fall within a safe level. EPA should consider the cumulative risk from all routes of exposure to a pesticide when setting tolerances. At a minimum, the Agency should factor drinking water exposure to pesticides into tolerances—especially because EPA now has substantial data on pesticide contamination of ground and surface water supplies.

Toxicologists know that in some instances exposure to a combination of toxic chemicals can result in a much higher incidence of cancers or other adverse health effects than would be expected from merely adding the predicted effects of each chemical. For example, smokers who were exposed to asbestos suffered far higher rates of lung cancer than were predicted by simply summing the risks of cigarette smoking and asbestos exposure. **Synergistic** interactions have been documented between organophosphate insecticides, such as disulfoton, and phosalone, methamidiphos and malathion, and EPN and malathion. Synergism and other non-additive effects of pesticides have not been adequately studied. EPA should require testing of pesticides which are commonly used in combination for synergistic effects.

In the absence of data demonstrating synergism, EPA should at least consistently assume that the effects of multiple pesticides are additive. In fact, the Agency has adopted guidelines for assessing the additive health risk from exposure to chemical mixtures. This approach is a good starting point for determining the health effect from simultaneous exposure to several pesticides. Although difficult to consider in establishing tolerances, synergism should not be dismissed by EPA as it could have serious health consequences. This is particularly true for exposure to organophosphates. EPA should explore methods to factor consideration of synergism into pesticide regulatory actions.

EPA Must Prohibit the Use of Dangerous "Inert" Ingredients in Pesticides.

EPA's current tolerance-setting procedure also ignores the risks posed by the so-called "inert" ingredients in the pesticide. Inerts are chemicals used to carry, dilute, or stabilize the "active" ingredients which effectively weaken, repel or kill the pest. Many of the "inerts" are dangerous chemicals in their own right. Carcinogenic "inerts" include asbestos, benzene, carbon tetrachloride, chloroform and formaldehyde. Neurological damage may occur from "inerts" such as hexachlorophene and lead. EPA's tolerances are based on a clearly erroneous presumption that the "inert" ingredient will not contribute to injury or illness. EPA should prohibit the use of dangerous "inert" ingredients.

EPA Must Require Pesticide Registrants to Develop Analytical Detection Methods Which Can be Routinely Used by the Government in Enforcing the Pesticide Laws. Such Methods Must be Swift, Accurate, and Feasible for the FDA Laboratories.

Pesticides should not be used on food if FDA cannot readily detect the residues in food. Without the ability to detect residues at reasonable cost, FDA cannot enforce EPA's tolerances and, therefore, cannot protect the public from pesticide residues in food.

EPA should require that registrants of food-use pesticides develop practical methods for detecting pesticide residues. An analytical method should be considered practical only if it can reliably and routinely quantify the level of residue in food with sensitivity sufficient to enforce the tolerance; if it can provide results in less than eight hours; if it can be conducted with existing FDA laboratory equipment; and if its use costs no more than what is typically incurred by the FDA in using a multi-residue method. A practical method could be an individual pesticide scan. However, broader multi-residue methods should be sought, where feasible.

Congress Must Clarify EPA's Authority to Revoke or Modify Tolerances Swiftly When Pesticides in Our Diet are Found to Present Significant Risk. EPA Needs the Authority to Revoke a Tolerance Automatically When Data Needed for Evaluating a Tolerance are Not Submitted According to EPA's Deadlines.

The food and drug law includes glacially slow, cumbersome procedures which EPA must follow when seeking to modify or revoke an old tolerance. Congress must revise these procedures. Even when pesticide makers fail to submit required toxicology data, EPA is unable to immediately halt the use of the chemical on food. Dr. John A. Moore, EPA Assistant Administrator, stated at last year that it was likely to take years to remove a pesticide from the market, even when the manufacturers flagrantly violate a data-submission requirement of the law. EPA's supposedly rapid Special Review process to investigate the hazards of dangerous pesticides takes many years. For example, the EBDCs have been undergoing special investigation since 1977.

EPA Must Require Systematic Testing of all Pesticides for Their Potential to Cause Neurological Damage.

EPA has not routinely required pesticide manufacturers to investigate the possible effects of pesticides on the nervous system. Even when EPA requires some study of the neurotoxic effects of a pesticide, it asks only for data which indicate at what levels the pesticide causes paralysis and death. Many chemicals are now known to have far more subtle but deleterious effects on the nervous system. For example, studies suggest that exposure to organophosphates and carbamate pesticides before and immediately after birth can cause delays in reflex and sexual development, delays in eye opening, alter nerve transmission function and neuroreceptor development, impair neuromuscular function, alter brain electrical activity, and affect brain structure. EPA needs to know what effects pesticide residues may have on motor coordination, memory, behavior, learning and intelligence. Neurotoxicity testing should be required for all pesticides used on food, and the testing should evaluate both acute and long-term adverse effects.

Legislation is Needed to Ensure That FDA's Program to Monitor for Pesticides in Food is Improved.

Food safety cannot be achieved by EPA reforms alone. The FDA has responsibility to monitor for most pesticides in food. For commodities other than meat, milk, poultry and eggs, for which the USDA has responsibility, FDA enforces the laws concerning food safety.

A minimum of six major FDA reforms is needed:

FDA Must Improve Its Methods for Detecting Pesticides in Food.

FDA monitoring for pesticides is inadequate to ensure that residues are legal, let alone safe. Of the 496 pesticides FDA has identified as likely to leave residues in food, FDA's routine analytical methods can only detect 203—only 41%. Of the 105 pesticides which FDA considers to pose a moderate to high health hazard, only 58 (55%) are detectible using the FDA multi-residue methods. Among the commonly used pesticides that cannot be detected by FDA's multi-residue methods are benomyl, daminozide, the EBDC fungicides and paraquat. Twenty-six of the 53 pesticides identified by EPA as potentially oncogenic for the 1987 NAS report on pesticides in food cannot be detected by FDA's multi-residue method.

Pesticide Use Information is Crucial to the Improvement of FDA's Pesticide Monitoring Program. Use information should be required to maintain a tolerance.

FDA, with the assistance of other federal agencies and growers, should develop accurate, detailed pesticide use information for both domestic and imported produce. FDA cannot effectively use single residue analytical methods to supplement the multi-residue pesticide screens because it has very limited information on pesticides used on each crop. FDA chemists would be wasting their time using monitoring methods capable of detecting only a single pesticide unless the government had fairly solid evidence that the pesticide, in fact, was applied to the crop during growing, storage or transportation.

Even with domestically grown crops, FDA lacks reliable pesticide application records. Often, FDA's choice of monitoring method must be based on educated guesses about what pesticides were likely to have been used. Some pesticide use information can be gleaned from USDA, state agencies, the food industry, and grower cooperatives, but the data vary significantly in comprehensiveness and reliability. According to the Resources for the Future, only nine states prepare pesticide usage reports on most crops, 17 states report only on major crops, and 15 other states prepare usage reports of more limited value.

With regard to imported foods, FDA has even less knowledge about the pesticides which may remain on them. FDA relies heavily on a private pesticide-use database which provides aggregate information for only 12 countries that export produce to the United States. Dozens of other countries ship significant amounts of fresh fruits and vegetables to the United States. Many pesticides that are banned in this country—such as DDT and EDB—are used in agriculture in countries from which we import substantial quantities of fruits and vegetables. Pesticides like omethoate are not registered for use in the U.S. but are used in other countries on produce which may be exported to this country. Additionally, pesticides that have limited legal uses in the U.S. may be used on entirely different crops abroad.

FDA Must Cut Delays in Analyzing Samples.

FDA must also decrease the time lag between sample collection and release of the results of the pesticide analysis. GAO found that between October 1, 1983 and March 31, 1985, FDA laboratories took an average of 28 calendar days to analyze samples for pesticide residues. Eighty-three percent of 179 illegal samples were not analyzed within the two-day period FDA said was usually needed for intercepting the food prior to sale to consumers. Congress should require FDA to increase the speed of its analyses of food samples, and to ensure that FDA can remove samples with illegal residues from the market before they are sold or consumed.

FDA Should Track the Disposition of Foods Bearing Illegal Pesticide Residues.

FDA should initiate a system to track the disposition of all food samples found to contain illegal pesticide residues. In 1986, GAO reported, "in 107 of the 179 cases in which FDA found that food contained illegal pesticide residues, it did not take any action to prevent food from reaching the market because the food had already been sold." According to GAO, in only one of the 179 identified incidents of illegal pesticide contamination did FDA seize all of the illegal food.

FDA should routinely collect and publicly report the Agency's disposition of food containing violative concentrations of pesticides. Such information is crucial for assessing FDA's effectiveness at preventing public exposure to unsafe levels of pesticides in food. Such reports would also help Congress judge the adequacy of FDA's sanctions against those responsible for violating pesticide tolerances.

Congress Should Give FDA Authority to Detain Domestic Food Shipments Pending Analysis and to Impose Civil Penalties for Violations of Pesticide Tolerances.

FDA should be given greater authority to control interstate marketing of food that contains illegal residues. This should include the right to initiate civil penalties against violators of the food safety laws, and the right to detain food suspected of containing residues pending results of laboratory analyses.

Currently, FDA can only initiate criminal penalties against violators. The severity of these sanctions and the heightened burden of proof requires the FDA to expend considerable resources and staff hours to support the charges. Consequently, FDA rarely takes legal action against violators. In the majority of cases an appropriate sanction would be significant civil penalties imposed against firms or individuals producing or shipping food with illegal residues in interstate commerce.

FDA also needs explicit authority to detain domestic food shipments suspected of containing illegal residues until the analytical results have been obtained. Without this authority, FDA frequently cannot prevent human dietary exposure to unsafe pesticide residues. Often, the contaminated food has already been consumed before final laboratory results are available.

FDA's Pesticide Monitoring Resources Must Be Enhanced.

FDA's monitoring program has suffered from substantial staff cuts during the Reagan Administration. In 1980, FDA allocated about $12.7 million and 344 staff years to monitoring for pesticides. In 1985, despite the greater dimensions of the problem, pesticide monitoring consumed $13.7 million and only 309 staff years. Resources available to the FDA's pesticide monitoring program need to be increased.

FDA Should Require Food Retailers to Label Fresh Produce With the Country of Origin. Interested Consumers Should be Able to Obtain Information on the Pesticides Used on Any Commodity Offered For Sale.

It is impossible for consumers to take action to limit their exposure to pesticides unless they can find out what pesticides were used in growing and storing the fruits and vegetables in their grocery stores. Consumers have a right to know about the pesticides used on foods they purchase. Grocery store managers should have information about the pesticides used in growing, storing and transporting the produce offered for sale. Food retailers could now require produce suppliers to provide information on pesticide use. Moreover, current federal law requires that pesticides applied to produce after harvest be identified on the shipping container. Concerned citizens should be able to review these types of information. In addition, the Federal Food, Drug and Cosmetic Act should be amended to include a provision giving consumers a right-to-know about pesticides residues in food. Under such a provision, FDA would be required to obtain data on pesticide usage on food crops and make this information available to the public on request.

Because uses of many pesticides which are illegal in the United States occur in other countries, consumers should be informed when they are purchasing produce from abroad. FDA statistics indicate that imported fruits and vegetables have a consistently higher rate of violation of the U.S. pesticide laws. In fact, between 1979 and 1985, the pesticide residue violation rate for domestically-grown commodities averaged 2.9% whereas the violation rate for imported commodities averaged 6.1%. If this trend continues, consumers may expect imported fresh produce to be twice as likely to violate the pesticide laws as domestic fruits and vegetables. FDA should require country-of-origin labelling so that consumers can immediately identify imported fruits and vegetables, which have become an increasingly large segment of the produce consumed in the U.S.

Congress Must Act to Assist Growers to Reduce Pesticide Residues.

In addition to more effective EPA and FDA programs to regulate pesticides, the food supply will be best protected if growers are provided incentives to reduce their use of pesticides. Many farmers are willing to consider techniques that decrease pesticide applications, but barriers exist to the adoption of these methods.

Six basic reforms in farm policy at minimum, are necessary to encourage growers to reduce pesticide use.

Congress Should Provide Credit Assistance, Crop Insurance, and Other Financial Protection for Growers Who are Changing From Conventional, Chemical-Intensive Agriculture to Innovative, Low-Residue Techniques.

Most farmers know pesticides are costly and dangerous. Nevertheless, many growers are reluctant to modify their farming practices because of the financial risks involved. Alternative techniques such as integrated pest management (IPM) and organic farming can be used to reduce greatly or even eliminate pesticide residues in many crops. Instead of routinely spraying fields with herbicides, insecticides, and fungicides, IPM growers monitor their fields for pest infestations and use control measures only when they determine that the pests have reached an economically damaging level. IPM and organic growers both rely on a variety of techniques to limit pest infestation, many of which do not entail the use of any chemicals. Generally, organic growers apply no synthetic pesticides. IPM growers may use some synthetic pesticides, but minimize use of chemicals they believe will disrupt the ecological balance of crop, pest and pest predators in their fields.

Although organic farming may ultimately prove profitable for many growers, there may be significant losses during the transition period in which the farmer learns and perfects the new techniques. Particularly for smaller growers who may not be able to survive several hard years, the disincentives to innovate are great.

Congress should assist growers who wish to modify their farming practices in ways that could greatly reduce pesticide use by offering them financial support. Credit assistance, crop insurance and other financial incentives should be provided for growers undergoing the transition from conventional farming to innovative, low-pesticide farming methods.

Congress Should Impose a Tax on Pesticides to Fund Demonstration of Farming Techniques That Will Result in Lower Pesticide Residues.

The development of alternative farming techniques to reduce pest problems, while reducing dependence on pesticides, requires research and field demonstration. In 1988, less than $4 million out of USDA's $1 billion annual research budget went toward the Department's new low-input ("sustainable agriculture") research program.

This valuable program and closely related endeavors should be greatly expanded. A tax as small as one percent on pesticide products could raise tens of millions of dollars in revenue. Congress should direct these funds to experiments demonstrating techniques that can reduce farmers' reliance on pesticides and thus limit pesticide residues in food. The 1990 Farm Bill provides an excellent opportunity for Congress to enhance research in IPM and organic growing techniques, funded in part by a tax on pesticide products.

Congress Should Establish National Definitions of "Integrated Pest Management" and "Organic" Farming Techniques

and Should Establish a National Certification Process for Commodities Grown Using These Techniques.

Consumers wishing to limit pesticide exposures may prefer to purchase products with lower residues grown using IPM or organic farming practices. However, not all states have defined "organic" farming, and those that have, are not always in agreement about the criteria which should apply. In the interest of consumer protection, there should be a national definition of "IPM" and "organic" farming, and a national certification process to ensure that produce identified as "IPM-grown" or "organic" is truly in compliance with the national definitions. Only if Congress adopts such a system will consumers know that purchasing fruits or vegetables certified as "IPM-grown" or "organic" provides the same degree of protection for Georgians and Vermonters as for Californians and Texans.

Congress Should Modify Federal Farm Support Programs to Ensure That Growers are Permitted and Encouraged to Use Crop Rotation and Other Pesticide-Reducing Techniques Without Jeopardizing Their Income.

Crop rotation can greatly reduce pest problems, particularly from soil-dwelling pests. However, many growers who qualify for support payments under the current federal commodity programs risk losing their eligibility, and hence their income, if they practice crop rotation. This anomaly of the law must be changed. Congress should ensure that growers do not jeopardize their farm program eligibility by employing prudent, pesticide-reducing farming practices, including rotation of annual crops.

Congress Should Ensure That Crop Insurance Policies Do Not Discriminate Against Growers Using Organic or Integrated Pest Management Farming Practices.

Some growers have reported that crop insurance is not available unless they promise to rely on pesticides. The practice of some lenders of making crop insurance available only to conventional growers creates a dilemma for growers who wish to develop alternative farm practices that could decrease pesticide residues in food. Growers should not be penalized for their innovations in reducing pesticide residues. Congress should ensure that federal crop insurance policies do not discriminate against IPM or organic food growers.

Congress Should Legislatively Modify Agricultural Supply-Control Systems to Ensure That They Do Not Demand Cosmetically Perfect Produce That Require Excessive Pesticide Use.

Currently, marketing and grading standards are developed by the agricultural industry to regulate the quantity of produce offered for sale in this country. Such standards normally require commercial fruit and vegetables to meet very detailed color, size, shape and texture criteria. Fruit or vegetables that may meet these rigorous cosmetic standards may be neither the best tasting nor the most nutritious produce. Supply-control systems should reward nutritionally exceptional produce, not produce whose "perfect" appearance is achievable only by reliance on pesticides. Consumers should be informed that uniform color or size is not ordinarily correlated with nutrition or taste. Congress must modify the supply-control system to ensure that growers are rewarded for food which is good to eat, thereby removing incentives for growing good-looking, but pesticide-laden food.

Some further resources

Breaking the Pesticide Habit: Alternatives to Twelve Hazardous Pesticides, Terry Gipps, International Alliance for Sustainable Agriculture, 1987. (Available from IASA, University of Minnesota, Newman Center, 1701 University Avenue SE, Minneapolis, MN 55414. $14.95.)

Describes alternatives to the pesticides included in Pesticide Action Network's "Dirty Dozen" campaign. This campaign identified 12 extremely hazardous pesticides that are responsible for most of the pesticide deaths and much of the environmental damage in the developing world.

Circle of Poison: Pesticides and People in a Hungry World, David Weir and Mark Schapiro, Institute for Food and Development Policy, 1981. (Available from IFDP, 145 Ninth St., San Francisco, CA 94103. $3.95.)

Documents the international sales of the most deadly pesticides and tells how they return to the United States via imported food.

Food and Drug Administration: Laboratory Analysis of Product Samples Needs to be More Timely, GAO, September 1986.

Pesticides: Better Sampling and Enforcement Needed on Imported Food, GAO, September 1986.

Pesticides: EPA's Formidable Task to Assess and Regulate Their Risks, GAO, April 1986.

Pesticides: Need to Enhance FDA's Ability to Protect the Public From Illegal Residues, GAO, October 1986. (Available from U.S. General Accounting Office, P.O. Box 6015, Gaithersburg, MD 20877. Free.)

Audits of the adequacy of most aspects of the federal government's program to regulate pesticides in food.

Guess What's Coming to Dinner: Contaminants in our Food, Americans for Safe Food, 1987. (Available from Center for Science in the Public Interest, 1501 Sixteenth St. NW, Washington, DC 20036. $3.50.)

An overview of many food contamination issues including pesticides, antibiotics, food additives, and PCBs.

Harvest of Unknowns: Pesticide Contamination in Imported Foods, Shelley Hearne, 1984. (Available from NRDC Publications Dept., 40 W. 20th St., New York, NY 10011. $7.50.)

A discussion of the unique problems posed by pesticides in imported food.

The Health Detective's Handbook: A Guide to the Investigation of Environmental Health Hazards by Nonprofessionals, M. Legator, B. Harper and M. Scott, 1985. (Available from The Johns Hopkins University Press, 701 W. 40th St., Baltimore, MD 21211.)

A thorough guide for citizens to evaluate whether adverse health effects in their community may be related to toxic substances.

Healthy Harvest II: A Directory of Sustainable Agriculture and Horticulture Organizations, S.J. Sanzone, J. Burman and M.A. Hage, eds., 1987–88.

Adapted with permission from *Pesticide Alert,* by L. Mott and K. Snyder, Sierra Club Books, 1987

(Available from Potomac Valley Press, 1424 16th St. NW, Suite 105, Washington, DC 20036. $10.95.)

Provides descriptions of the activities and publications of agricultural, research, and political organizations.

Intolerable Risk: Pesticides in our Children's Food, B. Sewell, R. Whyatt, J. Hathaway, and L. Mott, 1989. (Available from NRDC Publications Department, 40 West 20th Street, New York, NY 10011. $25. The Executive Summary of *Intolerable Risk* is available for $5.)

On the Trail of A Pesticide: A Guide to Learning About the Chemistry, Effects and Testing of Pesticides, Mary O'Brien, Northwest Coalition for Alternatives to Pesticides, 1984. (Available from NCAP, P.O. Box 1393, Eugene, OR 97440. $14.00.)

An exhaustive primer for citizen activists on pesticides, including discussions of toxicology and environmental effects.

Pesticide Alert: A Guide to Pesticides in Fruits and Vegetables, L. Mott and K. Snyder, Sierra Club Books, 1987. (Available from the Sierra Club Store, 730 Polk Street, San Francisco, CA 94109, 415-923-5500. $6.95 plus $3.00 postage and handling. California residents add applicable sales tax.)

Pesticides: A Community Action Guide, CONCERN, November 1987. (Available from CONCERN, Inc., 1794 Columbia Road NW, Washington, DC 20009. $3.00.)

An informative overview of pesticide issues.

The Pesticide Handbook: Profiles for Action, International Organization of Consumers Unions, Second Edition, 1986. (Available from International Organization of Consumers Unions, P.O. Box 1045, 10830 Penang, Malaysia.)

A description of the health effects of over 40 pesticides commonly used worldwide, and useful sources of information on international pesticide issues.

Pesticides in Food: What the Public Needs to Know, Lawrie Mott and Martha Broad, 1984.

Pesticide Reregistration: An Evaluation of EPA's Progresss, Lawrie Mott, 1986. (Available from NRDC, 90 New Montgomery St., San Francisco, CA 94105. $12.50 first class mail, $10.00 book rate.)

A 1984 analysis of the hazards posed by pesticides in food with recommendations for reform. A 1986 study of EPA's efforts to reevaluate the safety of older, inadequately tested pesticides.

Regulating Pesticides in Food: The Delaney Paradox, National Academy of Sciences, 1987. (Available from National Academy Press, 2101 Constitution Avenue NW, Washington, DC 20418. $19.95.)

A lengthy report on the potential risks posed by pesticides in food and different statutory approaches for regulating pesticides in food.

Silent Spring, Rachel Carson, Houghton Mifflin, Boston, 1962. (Available from Penguin Books, 625 Madison Avenue, New York, NY 10022.)

An articulate, readable, and still timely description of the problems posed by pesticides.

Pesticide reform organizations

Here is a brief list of the major organizations working on pesticide and related environmental issues. They may be able to respond to specific questions or provide information on technical, legal, or legislative issues. There are many other organizations actively involved in pesticide issues at the national, state, and local levels.

Americans for Safe Food
1501 Sixteenth Street, NW
Washington, DC 20036
(202) 332-9110

ASF is a coalition of over 40 consumer, environmental, and rural groups whose goal is to convert consumer dismay about health risks in food into progress toward the general availability of contaminant-free food.

National Coalition Against Misuse of Pesticides
530 Seventh Street, SE
Washington, DC 20003
(202) 543-5450

NCAMP is the primary national coalition of all grassroots organizations working on pesticide issues. Most pesticide reform organizations can be reached by contacting NCAMP. Five times a year, NCAMP publishes *Pesticides and You*, a newsletter on pesticide issues across the nation.

Natural Resources Defense Council
90 New Montgomery Street
San Francisco, CA 94105
(415) 777-0220

NRDC is dedicated to the protection of public health and the environment, and has worked extensively on seeking effective control of pesticides.

Northwest Coalition for Alternatives to Pesticides
P.O. Box 1393
Eugene, OR 97440
(503) 344-5044

NCAP, a five-state coalition of citizen groups working on pesticide issues, publishes the quarterly *Journal of Pesticide Reform*, a comprehensive publication on a wide variety of pesticide topics and related concerns.

Pesticides Action Network—North America Regional Center Pesticide Education Action Project
P.O. Box 610
San Francisco, CA 94101
(415) 771-7327

PAN is an international coalition of environmental, consumer, farmer, and research organizations, voluntary development agencies and individuals who are opposed to the worldwide misuse of poisonous pesticides.

Organizations promoting sustainable agriculture

The majority of organizations involved in sustainable or alternative agriculture are active on the regional or local level. Here are some of the groups with a national focus that can identify the most suitable group in your local area.

Bio-Integral Resource Center
P.O. Box 7414
Berkeley, CA 94707

BIRC specializes in information on least-toxic methods for managing any pests found in homes or gardens. *The Common Sense Pest Control Quarterly* provides up-to-date information about pests found in homes, on pets, and in the garden, and is written for non-technical audiences. A Publications Catalog which lists about 50 publications on individual pests is available for $1.00.

Organic Crop Improvement Association
P.O. Box 729A
White Oak Road
New Holland, PA 17557

OCIA is a farmer-owned international organization which provides information on methods to reduce and eliminate pesticide use and certifies individual farmers that meet OCIA's organic farming standards. If you send a self-addressed stamped envelope, they will send you a list of sources of OCIA-certified organic food in your area.

Organic Food Producers Association of North America
P.O. Box 31
Belchertown, MA 01007

OFPANA is a trade association formed by organic farm associations, food processors, distributors, and supporters to establish and maintain standards of excellence for organic food businesses. OFPANA is a marketing network for high-quality, authentic organic food that should be contacted by supermarkets seeking a supply of organic food.

Rodale Press and Research Center
33 E. Minor Street
Emmaus, PA 18098
(215) 967-5171

The Rodale Center publishes *Organic Gardening, New Farm,* and *Prevention* magazines. The Center also develops, tests, and publicizes new farming techniques that reduce the need for pesticides and chemical fertilizers. If you send a self-addressed stamped envelope to *Organic Gardening's* Reader Service, they can provide you with free lists of organic farming certification organizations, advice and information groups, a directory of resources on alternative agriculture, and an information packet on organic pest control—all developed for their Regeneration Gardeners Network.

Glossary

Active ingredient. The substance in a pesticide product designed to kill or control the target organism. Other ingredients in pesticide products, called "inerts," do not affect the target organism.

Acute toxicity. The toxic reaction that usually occurs shortly after exposure to a toxic agent (*e.g.*, a few hours or days).

Cancer. The unregulated overgrowth of cells. In medical terminology, a cancer is a malignant tumor.

Carcinogen. A substance that can produce cancer (malignant tumors) in experimental animals or is known to do so in humans.

Fungicides. Chemicals used to kill or suppress the growth of all fungi or a certain fungus.

Growth regulator. A preparation which alters the behavior of plants or the produce thereof through physiological (hormonal), rather than physical, action. It may act to accelerate or retard growth, prolong or break a dormant condition, or promote rooting.

Herbicides. Chemicals used to kill or suppress the growth of all or a certain type of plant.

Illegal residue. The presence of an active ingredient in amounts above the tolerance on a crop at harvest. In some cases, any amount of chemical present on the crop is considered illegal if no tolerance exists for the pesticide on the commodity.

Inert ingredient. A substance contained in a pesticide product or formulation that is not intended to kill or control the target pest. Materials include solvents, emulsifiers, wetting agents, carriers, diluents, and conditioning agents.

Insecticides. Chemicals used to kill a wide variety or a specific type of insect.

Integrated pest management. The use of two or more methods to control or prevent damage by a pest or pests. These include cultural practices, use of biological control agents, and can even include the use of selective pesticides.

Metabolites. A compound derived, in the case of a pesticide, by chemical, biological, or physical action upon the pesticide within a living organism (plant, insect, or higher animal). The metabolite may be more, equally or less toxic than the original compound. Metabolites can also be produced by the action of environmental factors such as temperature or sunlight.

Mutagen. A substance or agent that produces genetic changes in living cells.

Neurotoxicity. The state of toxic effects on the nervous system. Severe neurotoxic effects can include visual problems, muscle twitching and weakness, and abnormalities of brain function and behavior.

Organochlorines. A class of chemical compounds produced by the addition of chlorine atoms to hydrocarbons. Many of them (*e.g.*, DDT, dieldrin, and endrin) had insecticidal properties and became the most successful of

Adapted with permission from *Pesticide Alert*, by L. Mott and K. Snyder, Sierra Club Books, 1987.

the early synthetic insecticides. These insecticides are characterized by their persistence in the environment.

Organophosphates. A class of pesticides containing phosphorus that are used primarily as insecticides by disrupting nerve function.

Persistent pesticides. Pesticides that remain in the environment and do not degrade or metabolize to innocuous constituents for months or perhaps years.

Residue. That quantity of a substance, especially of pesticide active ingredient, remaining on or in a surface or crop (including livestock products).

Rodenticide. A preparation intended for the control of rodents (rats or mice) and related animals, such as gophers.

Special review. A regulatory procedure adopted by the U.S. EPA to rapidly review the hazards of a pesticide in order to decide whether the chemical should remain in use and whether any restrictions should be placed on its use. A chemical is placed into special review, formerly called Rebuttable Presumption Against Registration (RPAR), when its risks exceed certain criteria EPA has established, such as causing cancer or environmental harm.

Synergism. The tendency of chemicals acting in combination to produce effects greater than the sum of the effects of the individual chemicals.

Tolerance. The maximum amount of pesticide residue that is legally permitted in a food. EPA sets a distinct residue limit for each individual food to which the pesticide may be applied.

Toxicity. The harmful effects produced by a chemical.